The Fifth Generation
AN ANNOTATED BIBLIOGRAPHY

Max & Dawn Bramer

Addison-Wesley Publishing Company
Wokingham, England • Reading, Massachusetts • Menlo Park, California
Don Mills, Ontario • Amsterdam • Sydney • Singapore • Tokyo
Mexico City • Bogota • Santiago • San Juan

© 1984 by Addison-Wesley Publishers Limited

All rights reserved. No part of this publication may be reproduced, stored in a retrieval system, or transmitted in any form or by any means, electronic, mechanical, photocopying, recording, or otherwise, without prior written permission of the publisher.

Typeset by Blackmore Press, Shaftesbury

Cover design by Design & Graphic Art

Printed in Finland by Werner Söderström Osakeyhtiö, Member of Finnprint.

British Library Cataloguing in Publication Data

The Fifth generation: an annotated bibliography.
   1. Computers — Bibliography
   I. Bramer, M. A.   II. Bramer, Dawn
   016'.00164   Z5640

Library of Congress Cataloging in Publication Data

Bramer, M. A. (Max A.), 1948–
   The fifth generation.

   Includes index.
   1. Computers — Bibliography. 2. Computers — Research — Japan — Bibliography. I. Bramer, Dawn. II. Title.
   Z5640.B73   1984   [QA76]   016.00164   84-19135

   ISBN 0-201-14427-1

ABCDEF 8987654

The Fifth Generation
AN ANNOTATED BIBLIOGRAPHY

To Frances *and* Bryony
— *who will view the Fifth Generation as history*

Contents

Japan and the Fifth Generation   1

Bibliography

A   Principal Documents   12
B   Overview and Background Papers   18
C   The Human-Computer Interface   36
D   Parallel Processing, Novel Architecture and VLSI   44
E   Logic and Functional Programming   64
F   Expert Systems and Artificial Intelligence   78
G   Networks   104

Reference Sources   106

Author Index   111

Subject Index   115

Preface

The Japanese announcement of a Fifth Generation Computer Project towards the end of 1981 caused consternation amongst computer users and computer professionals alike.

The Japanese proposals constitute a radical re-appraisal of the functions which an advanced computer system should be able to perform, the programming languages needed to implement such functions, and the machine architectures suitable for supporting the chosen languages.

A Fifth Generation system is envisaged as a series of interconnected database and parallel processing machines, accessed by means of an 'intelligent interface machine' which can accept problem statements in natural language, either in typed form or as continuous speech. The information it processes will not be data in the conventional sense, but knowledge. The Japanese JIPDEC report describes Fifth Generation computers as 'knowledge information processing systems' in which 'intelligence will be greatly improved to approach that of a human being'.

Although the key components of the Fifth Generation systems (such as VLSI technology, parallel machine architectures, logic programming and applied Artificial Intelligence, in particular Expert Systems) have been pioneered in the West, relatively few (even amongst computer professionals) have any experience or knowledge of the techniques involved.

This book aims to guide both computer users and computer professionals through the ever-growing literature on the Japanese project and the unfolding national and international responses (such as the British Government's Alvey programme and the Microelectronics and Computer Technology Corporation (MCC) in the USA).

It comprises an introductory chapter, which describes and assesses the Fifth Generation proposals and their implications, followed by a bibliography, which provides abstracts of a selection of around 250 of the most important English-language publications in the field (including translated items), up to the end of 1983.

These include unpublished reports as well as books and journal articles, and material at a general policy or 'overview' level as well as more detailed technical articles.

As well as many items on the Japanese proposals themselves, and reactions to them from Britain and the US, the bibliography contains a number of items on work relating to the individual components identified above.

As well as many items on the Japanese proposals themselves, and reactions to them from Britain and the US, the bibliography contains a number of items on work relating to the individual components identified above. The bibliography is divided into seven sections:

- Part A contains 'principal documents' such as the proceedings of the

international conference in Tokyo in October 1981 which launched the Japanese programme.
- Part B contains 'overview and background papers', together with items evaluating the fifth generation proposals and assessing their likelihood of success.
- Parts C to G contain papers on specific fifth generation topics, namely 'the human-computer interface', 'parallel processing, novel architectures and VLSI', 'logic and functional programming', 'expert systems and artificial intelligence' and 'networks'.

In all cases full bibliographical details are given and each abstract includes a list of keywords which link it to a subject index. An index to all authors is also included.

A section on reference sources gives the full addresses of publishers and other sources from which all documents included in the bibliography may be obtained.

Although only English language material has been included in the bibliography, translations of papers from the Japanese project feature substantially in it.

No attempt has been made to be exhaustive; rather, effort has been concentrated on selecting the most useful items to serve as an entry into the field for the general as well as the specialist reader. The aim has been to provide enough information to enable the reader to decide whether study of the original would be worthwhile.

Priority has been given to recent publications and to publications which survey and analyse the Fifth Generation project and its hardware and software components.

It is not intended to replace specialist sources of information for 'leading-edge' researchers in the field, although it is hoped that they will also find it of value. However, very detailed technical material has been omitted, particularly on VLSI. Papers on specific user applications have generally been omitted also.

Although easily available sources of information (such as books and journals) have been given preference, less readily available items have also been included when particularly significant (with full details of the sources from which they can be obtained).

Overall, it is hoped that the bibliography provides enough information to facilitate a study of Fifth Generation computers as currently envisaged and of their technical context.

The authors would like to thank the reviewers, especially Philip Treleaven and Bob Muller, for their valuable comments. The introductory chapter 'Japan and the Fifth Generation' is reproduced with permission from *New Information Technology* edited by A. Burns (1984), and published by Ellis Horwood Limited, Chichester, England. Figures 1 and 2 are taken from 'The JIPDEC report' and are reproduced courtesy of the Japan Information Processing Development Center (JIPDEC).

Japan and the Fifth Generation

It is customary to divide the history of computing into four generations, based largely on the type of component used in the machines' construction. There are no rigidly defined boundaries but, in general terms, the first generation ran from 1945 to the mid 1950s and included the early valve computers such as EDSAC, the Electronic Delay Storage Automatic Calculator, constructed at Cambridge in the late 1940s. The first generation was superseded by a second generation based on the use of transistors, which lasted until about 1965. The third generation utilised integrated circuits, containing a number of components on a single chip of semiconductor material.

There is less general agreement about what constitutes a fourth generation computer. Some apply the term to machines making use of large scale integrated circuits (LSI), which appeared in the early 1970s, others confine it to machines using very large scale integrated circuit (VLSI) technology, which are only now becoming available. Although the technology of fourth generation systems is superficially more sophisticated than that of the first three, the underlying architecture is still that of the 'von Neumann' computers, based in particular on the sequential processing of instructions, and the principal programming languages (such as Cobol or Fortran) remain the traditional ones of earlier generations.

What will ultimately replace the fourth generation systems is, of course, largely a matter for speculation. The Japanese Fifth Generation Computer Systems proposals are a bold attempt to specify what the next generation should comprise, looking up to a decade ahead, and then to bring that vision about by means of a well-funded, nationally co-ordinated research and development programme in collaboration with other countries.

The most striking feature of the fifth generation proposals is that they do not represent a natural progression from the first four generations, based on increasingly sophisticated electronics, but rather constitute a radical reappraisal of the functions which an advanced computer system should be able to perform, the programming languages necessary to implement such functions and, finally, what machine architecture is suitable for supporting the chosen languages. If successful, the outcome of this initiative will be fifth generation systems which are not a continuing development of the previous four or likely to be compatible with them, but which are demonstrably far more powerful than their commercial competitors. It is perhaps not surprising, then, that the Japanese proposals have been widely interpreted as an attempt by Japan to bypass the computer technology of the West and become the world's leading supplier of advanced computer systems within ten years.

The Japanese Proposals

The Japanese Fifth Generation Computer project was announced at an international conference in Tokyo in October 1981, attended by senior representatives of most of the Western computer industries, and caused an immediate sensation. The proposals were the outcome of two years of study and research organised by JIPDEC (the Japan Information Processing DEvelopment Center) and were published as the 'Preliminary Report on Study and Research on Fifth-Generation Computers', more popularly known as 'The JIPDEC Report' (JIPDEC, 1981). The proposed programme of research and development aims at bringing together a number of sophisticated ideas and projects currently being independently developed and combining them into a composite 'fifth-generation system', by the target year of 1990. In broad terms, a fifth generation system is envisaged as a series of interconnected database and parallel processing machines, accessed by means of an 'intelligent interface machine' which can accept problem statements in natural language, either in typed form or as continuous speech.

Even this ambitious statement obscures the fact that the information processed by a fifth generation system is expected to be not data, in the usual sense of numbers, names and addresses and so on, but knowledge. The JIPDEC report states that:

> The Fifth Generation Computer Systems will be *knowledge information processing* systems having problem-solving functions of a very high level. In these systems, *intelligence will be greatly improved to approach that of a human being.* (Authors' italics)

It is the phrase 'intelligence will be greatly improved to approach that of a human being' which more than any other has caused alarm — and also not a small amount of scepticism — in the West. The major application area for the information processing of the 1990s is expected to be the solution of complex problems which involve (for people) reasoning and inference.

What JIPDEC calls Knowledge Information Processing is the province of the branch of Computer Science known as Artificial Intelligence. Problem-solving systems of the kind generally referred to as Expert Systems are envisaged as a major field for the information processing of the 1990s and computers designed for this purpose are the principal theme of the fifth generation computer project.

Structure of a Fifth Generation Computer

Figure 1 shows what the JIPDEC report calls the 'basic configuration image' of fifth generation systems.

The report comments thus:

> The machines are to be structured according to function on various new architectures, including a data flow machine, which are based on VLSI architecture and each system is to be a combination of machines suitable for various individual applications or needs.
>
> Furthermore, from a macro configuration point of view, having the system shown in the figure as one of the principal elements, a multiple system form of usage where this would be connected to a local or global network and the whole network then be utilized as a large-scale distributed processing system, is also being envisioned.

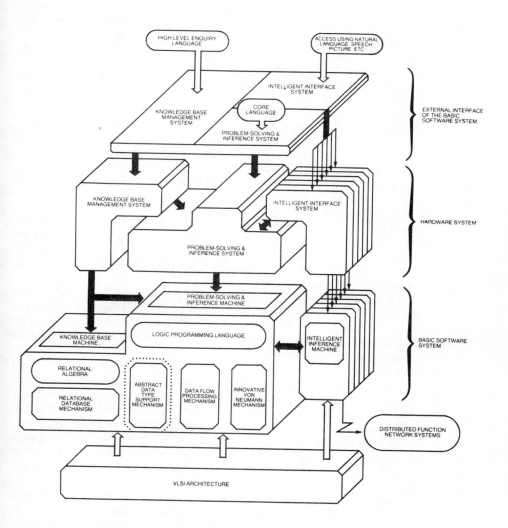

Figure 1. Basic configuration image of fifth generation computer systems

Although, like much of the JIPDEC report, Figure 1 is difficult to comprehend in detail, the outline is reasonably clear and it is difficult not to be impressed by its ambitious nature.

The 'basic software system' is shown as having three components: a 'problem-solving and inference system', supported by a 'knowledge base management system' and an 'intelligent interface system'. Access to the system using natural language, speech, pictures, and so on is indicated in the figure. Considerable importance is attached to the ease of use of fifth generation systems, which are expected to be able to handle input and output via voice, graphics, images, or documents. Machine translation is also seen as an important need. These complex tasks are handled by the 'Intelligent Interface System'.

As indicated previously, one of the major functions of a fifth generation system will be to solve problems by reasoning and inference in much the same way as people do. This includes the capability to make judgements from inexact or incomplete information and to take part in a question-and-answer dialogue with the user (such as a medical consultation with a patient or a physician). This task is handled by the 'Problem-solving and Inference System.

The core programming language for this system is envisaged to be a substantially extended form of the 'logic programming' language Prolog, which is based on the use of such rules as:

if A, B and C are true then deduce D

supplemented by a built-in theorem prover to form valid deductions from the given rules. This 'declarative' form of programming is radically different from that of conventional languages of the 'procedural' kind, and arises from research into Automatic Theorem Proving, largely by the Artificial Intelligence community in Britain and Europe.

The third component of the basic software system is the 'Knowledge Base Management System'. This is an extended form of the data-base management systems that exist today, with the difference that the information handled will be much closer to human knowledge than the data in conventional databases. It might, for example, take the form of rules of the kind above.

Each of the three components of the basic software system has its own correspondingly specialised machine, based upon Very Large Scale Integration (VLSI) architecture. Thus the hardware is shown as comprising a 'knowledge base machine', a 'problem-solving and inference machine' and an 'intelligent interface machine'. It is envisaged as having a highly parallel architecture, possibly based on dataflow, as a means of achieving much higher execution speeds than contemporary computers. Finally, it is expected that the fifth generation computers will range in size from small (personal) to large (mainframes) and will be interconnected via local and global networks.

To summarise, the key components of the Japanese proposals would seem to be VLSI technology, advanced parallel machine architectures (such as dataflow), logic programming and 'applied Artificial Intelligence' (especially Expert Systems). Although each of these parts have been pioneered and developed in the West, they are far from being fully exploited — or, in some cases, even widely understood — at present. The achievement of the Japanese project is to bring them together as a unified vision of what is required in a computer system of the 1990s.

The project comprises four stages: preliminary, initial, middle and final. The preliminary stage (information gathering) lasted from 1979–1981. The initial stage began in April 1982 and will last for three years, during which the basic technology of the fifth generation will be developed. The target of the middle stage (1985–1988) is the development of subsystems and possibly a prototype fifth generation system. The goal of the final stage (1989–1991) is the construction of a complete fifth generation system.

Research and Development Themes for the Fifth Generation

JIPDEC has identified 26 research and development themes for fifth generation systems, grouped into 7 categories. These are shown in Figure 2, which is taken from the JIPDEC report.

Basic application systems	1-1)	Machine translation system
	1-2)	Question answering system
	1-3)	Applied speech understanding system
	1-4)	Applied picture and image understanding system
	1-5)	Applied problem solving system
Basic software systems	2-1)	Knowledge base management system
	2-2)	Problem solving and inference system
	2-3)	Intelligent interface system
New advanced architecture	3-1)	Logic programming machine
	3-2)	Functional machine
	3-3)	Relational algebra machine
	3-4)	Abstract data type support machine
	3-5)	Data flow machine
	3-6)	Innovative von Neumann machine
Distributed function architecture	4-1)	Distributed function architecture
	4-2)	Network architecture
	4-3)	Data base machine
	4-4)	High-speed numerical computation machine
	4-5)	High-level man-machine communication system
VLSI technology	5-1)	VLSI architecture
	5-2)	Intelligent VLSI CAD system
Systematization technology	6-1)	Intelligent programming system
	6-2)	Knowledge base design system
	6-3)	Systematization technology for computer architecture
	6-4)	Data base and distributed data base system
Development supporting technology	7-1)	Development support system

Figure 2. Research and development themes for fifth generation computer systems.

Each theme has its own target specifications. Thus, the machine translation system (theme 1-1) must ultimately be able to handle 100,000 words with ninety percent accuracy at a cost of thirty percent or less of that of human translators. The problem-solving and inference system (theme 2-2), which is central to fifth generation systems, must have the ability to perform from 100 to 1,000 million lips. 'Lips' stands for 'Logical Inferences Per Second', where a logical inference in this sense is estimated as the equivalent of 100 to 1,000 conventional machine instructions. Present day computers operate at a mere 10,000 to 100,000 lips.

The specification for the intelligent interface system (theme 2-3) is no less impressive: it must be able to accept speech input, from different speakers who are not nominated in advance, working almost on a real-time basis, and must be able to output speech in both Japanese and English. The target for VLSI architecture development (theme 5-1) is ten million transistors per chip, with the design of these integrated circuits also automated. These examples are enough to give the flavour of the research and development targets. By any standards, they are ambitious to the point of being revolutionary.

The JIPDEC report's frequent use of terms such as 'knowledge', 'intelligence', 'problem-solving', 'inference', 'learning', 'intelligent interface system', 'access using natural language', 'intelligent programming system', 'innovative von Neumann machine', 'abstract data type support machine', together with its astonishing performance targets give an impression of a project of the most fundamental importance. Much of the report would not have been taken seriously if it were put forward by a university research team looking for funds. What stops the JIPDEC report being regarded as merely a catalogue of good ideas is the simple fact that it is being presented as a serious research and development project organised on a national scale with a reputed funding of £250 million over ten years and a target date of 1990. Moreover, it is backed by a group consisting of senior members of the business and industrial community of a nation with an outstanding record of success in other technological fields. It is against this background of success that the fifth generation proposals need to be judged.

The Japanese Dimension

The remarkable success of the Japanese economy since the Second World War is already very well known in outline, not least in comparison with the severe decline of the economy of the United Kingdom. What is perhaps the most remarkable aspect of this success is that it has been achieved despite an almost total lack of material resources, compared with Britain's huge reserves of oil and coal, and with a steadily ageing population. The Japanese success is based upon the development and exploitation of advanced technology, in particular by making skilled use of the 'copy and improve' technique. In this it is clear that it is people who are regarded as the key national resource. The Japanese workforce is probably the most highly educated in the world: ninety-four percent of children attend school to the age of eighteen, compared with twenty-two percent in the UK* and thirty-seven and a half percent subsequently enter higher education, as against twelve percent in the UK. Whereas the funding of UK universities has been cut drastically in recent years, the declared long-term aim of the Japanese government is to provide university education for every child.

*The figures here and elsewhere in this section are drawn from an excellent report by Dr. F. G. Marshall, Counsellor for Science and Technology at the British Embassy in Tokyo, entitled "Japan — Prosperity from Technology".

Although Japanese industry is highly automated, with an estimated seventy-five thousand 'fixed sequence' robots and around ten thousand reprogrammable ones (the US has three thousand of the latter, the UK less than two hundred), the unemployment level is remarkably low: a mere two percent at November 1980, compared with seven percent in the UK and seven point six percent in the US. This is due to the national policy of 'lifetime employment' by the same company. Workers whose jobs are automated are retrained for a more skilled and responsible job. Inevitably, companies tend to diversify their interests as more and more processes are automated.

The Japanese population would appear to be a highly contented one. It is perhaps indicative of this that surveys show that over ninety-five percent of Japanese regard themselves as middle-class.

For cultural reasons, as much as any others, it is possible for government agencies to launch national projects involving close collaboration between companies, in a way which seems inconceivable in Britain. For comparison with the fifth generation project, the other current national plans of the Agency for Industrial Science and Technology (AIST), part of the Ministry for International Trade and Industry (MITI), are listed below*. The companies concerned receive non-recoverable one hundred percent government grants.

Project	£million	Period
Post Robotics	28	1977–1983
Optoelectronics	42.8	1979–1986
Carbon Chemistry	35.7	1980–1986
Jet Engines	47.4	1971–1981
Steel	33.3	1973–1980
Oil Chemistry	32.8	1975–1981
Resource Recovery	26.2	1976–1981
Subsea Oil	35.7	1978–1984

This table places the fifth generation project in its national context. The total annual funding of all the AIST projects is £260 million. The fifth generation project must also be viewed in the light of the declared objective for the Japanese computer industry to match IBM in world-wide sales value. It would be easy to dismiss such an ambition, but it must be borne in mind that at least three Japanese computer manufacturers now claim that their machines are more powerful than those of IBM and, in 1980, both Fujitsu and Hitachi sold more computers in Japan (by value) than IBM.

The previously mentioned point that fifth generation systems would not be upward compatible with those of the previous four generations, especially by virtue of the 'intelligent' knowledge processing functions, is by no means unimportant. If fifth generation systems do eventually appear in the envisaged form, Japan's industrial competitors will suddenly find themselves in the unenviable position of selling computers that are not only demonstrably less

---

*Again, taken from "Japan — Prosperity from Technology", taking one pound sterling as equal to 420 yen.

sophisticated but also incompatible with the Japanese machines. It is not surprising, then, that interest in the US and Europe has been so intense.

The innovative nature of the fifth generation proposals has been remarked upon by many observers as markedly at variance with the standard Japanese technique of 'copy and improve'. It would seem that, in this field at least, 'copy and improve' is giving way to a serious attempt to take a technological lead based on innovation within Japan. This change of policy by a nation with such a record of technological success can only be regarded as a serious threat to the computer industries of the West. If the stated performance targets prove at all realistic, the fears of many western commentators may very well prove justified.

The Artificial Intelligence Connection

Although the hardware and machine architecture side of the Fifth Generation programme (VLSI, parallel architectures, distributed processing) is ambitious, it does at least build on 'mainline' developments elsewhere. For example, a dataflow machine is under construction at Manchester University which substantially predates the Japanese proposals. It is the software and applications side of the proposals which is simultaneously the more exciting and the more speculative. The idea of processing knowledge rather than data dominates the JIPDEC report. Fifth generation systems will be applied to the solution of complex problems, making use of stored knowledge bases for specialised fields and including facilities for automatic learning and inference. Such application systems are today generally known as Expert Systems and are the subject of considerable attention in the United States and Britain at the moment.

Although some have assumed that Expert Systems are a new phenomenon, it is only the name which is new. The earliest acknowledged Expert System — DENDRAL — dates back to 1965. DENDRAL was designed for use by organic chemists to infer the molecular structure of complex organic compounds from their chemical formulas and mass spectrograms. (Mass spectrograms are essentially bar plots of fragment masses against the relative frequency of fragments at each mass.) To do this, DENDRAL makes use of rules which relate physical features of the spectrogram (high peaks, absence of peaks, etc.) to the need for particular substructures to be present in or absent from the unknown chemical structure. The program is claimed to rival expert human performance for a number of molecular families.

Although DENDRAL is the oldest Expert System, in a number of ways it is a less 'typical' one than MYCIN, a system which diagnoses the cause of certain bacterial infections (especially in the blood) and recommends appropriate drug treatment, on the basis of an interactive dialogue with a physician about a particular case. To do this, MYCIN makes use of a 'knowledge base' of over four hundred rules of the form:

IF [condition] THEN [implication]

each of which has an associated 'degree of certainty' or 'certainty factor' indicating a subject expert's level of confidence in the rule. Thus, the following is a MYCIN rule (Shortliffe, 1976):

IF:

1) The site of the culture is blood, and
2) The gram stain of the organism is gramneg, and
3) The morphology of the organism is rod, and

4) The patient is a compromised host
THEN:
There is suggestive evidence (0.6) that the identity of the organism is pseudo-aeruginosa.

The use of rules of this kind as a means of representing complex knowledge is typical of Expert Systems, or at least of the kind so far developed in the West. Not only do such rules enable inexact (or heuristic) information to be included, but they are also represented and stored in memory in an explicit form which is thus capable of being studied — and, if necessary, changed — by a subject expert, in a fashion which is simply not possible with a conventional program in, for example, FORTRAN or COBOL.

As well as a knowledge base of rules, an expert system has a so-called inference engine which manipulates the rules to form valid conclusions, diagnoses and so on. In principle, the inference engine can be made independent of the rules included in the knowledge base, so that to construct a new expert system involves specifying a new set of rules but not the surrounding framework into which they fit. In addition to the ability to solve problems, expert systems are generally able to explain, and even discuss at length, their own chains of reasoning; for example, what rules acting on what clinical data led to a particular diagnosis. Again, the explicit encoding of knowledge in rules makes an English-language explanation possible in a way that is virtually impossible for a program in a conventional computer language.

At the present time, there are around forty expert systems in existence, primarily in the fields of medicine, science and engineering (Bramer, 1982). There are very few however in regular use in the field. It would seem that there is a widespread suspicion of 'intelligent' systems by business and industry, as well as by Government, in the West. It is indeed interesting to note that the Japanese, with virtually no expertise in the field, have placed such systems in a leading role for their national research and development programme for fifth generation systems.

Although Expert Systems, also known as Intelligent Knowledge-Based Systems, are a most important manifestation of Artificial Intelligence concerns in the Japanese proposals, they are certainly not the only ones. The development of an intelligent user interface, for instance the ability to understand continuous speech input, is an Artificial Intelligence task. Whereas the acceptance of a small number of clearly differentiated single word commands (STOP, GO, LEFT, RIGHT, and so on) is perfectly possible now, continuous speech is a very much harder proposition. Such speech often consists of a string of incomplete utterances with little apparent coherence. Even a single sentence can often take considerable prior knowledge to understand. The command "Pick it up!", for example, requires a knowledge of what is meant by the word "it". Differentiating a word from one that follows is also difficult without a good understanding of common sentence constructions and phrases. Thus the development of an intelligent user interface requires the use of stored knowledge about words and sentences (and the world), far more than improved technology. For comparison, IBM is reported to have a system which transcribes conversations from speech to printed text, based upon a pre-defined vocabulary of one thousand words. Even this performance, which is extremely modest in the face of the Japanese proposals, has been greeted with much disbelief.

The emphasis on speech understanding, language understanding and even machine translation from Japanese to English, or vice versa, again aligns the Fifth Generation project with work in Artificial Intelligence. Success in this sphere depends on theoretical advances, which are inevitably less predictable than, say, advances in miniaturisation technology. On the Knowledge Information Processing or Expert System front, it may only be necessary to apply the standard techniques which are already known, albeit only by a small band of devotees, in the West. The plans for intelligent user interfaces go far beyond what is currently possible and may not be attainable within the lifetime of the project. There are, however, research workers actively studying each separate topic and it would be foolish to assume that they will fail.

The third Artificial Intelligence aspect of the fifth generation proposals is the use of logic programming as the 'core' language. This has been widely misreported as meaning the language Prolog but, while this will be taken as the starting point, it is not the final solution. Prolog (Clocksin and Mellish, 1981) is essentially a rule-based programming language, in which a rule, such as the earlier one quoted for MYCIN, can be directly encoded in a single statement. The method used for combining rules is rather restrictive, however, and the language has a number of defects for more complex applications (whilst still remaining easier to use than most others for such applications). The intention is for fifth generation systems to use a much improved and embellished version of Prolog. There seems to be no substantial reason why this should not successfully be accomplished and the resulting language would certainly represent a major step forward if it became widely accepted as a general-purpose language for scientific and commercial computing.

In summary, although the envisaged fifth generation is not a continuous development from the previous four, there is much substance to the claim that it arises naturally from existing research into Artificial Intelligence. The difference is that, whereas the Japanese project will be well-funded and nationally organised and supported, Artificial Intelligence work in the West is usually carried out in small and badly-funded research groups, especially in universities, with little or no national co-ordination. It is possible that, largely as a result of the Japanese proposals, this position may now change.

The West's Response

The publication of the Japanese proposals has led to considerable interest and activity in the West. One valuable side-effect has been to focus attention on previously neglected work, especially on dataflow architectures, logic programming and expert systems. It is only fair to report, however, that the proposals have attracted not only interest but also scepticism from a number of quarters. Manufacturers of traditional systems frequently have little or no understanding of the background to the proposals and regard them as fanciful. Even amongst researchers in the areas concerned there has been much scepticism about the performance targets and the timescale for their achievement. One United States critic is quoted as saying:

> The Japanese will not dominate the next generation even with help from intellectual dilettantes who hate IBM blindly. They have an outstanding ability for low-cost, high-volume production, but do not appear to have a great receptivity to new ideas or aptitude for high technology. It is extremely difficult to name a single major idea in computing, hardware or applications, that came from a Japanese source.

A more positive opinion is held by Edward Feigenbaum, head of the Heuristic Programming Project at Stanford University (home of DENDRAL, MYCIN, and many other expert systems). Feigenbaum is quoted as stating, at a recent conference, that the Japanese have correctly spotted the beginning of the age of machines which engage in symbolic manipulation and not numeric computation. Even if they achieved only twenty percent of their goals, they would still be 'ahead of the game'. As long as they were on the right course, it would not even matter if they missed their time goals by orders of magnitude, they would still have a major lead on American companies.

His views seem to be shared by many in the United States, where the Microelectronics and Computer Technology Corporation (MCC) has been established as one response to the Japanese Fifth Generation project. This joint venture corporation funds research and contributes researchers in four major areas: computer-aided design and manufacturing, computer architecture, software technology and packaging.

In the United Kingdom, the government has accepted a report commissioned in 1982 by the Department of Industry from a committee chaired by Mr John Alvey, senior director of technology at British Telecom (the Alvey Committee), which recommended, as a response to the Japanese proposals, a national programme for Advanced Information Technology costing £350 million over five years, with a government contribution of two-thirds of that sum, the rest being provided by industry (Alvey Committee, 1982).

The Alvey Programme is now under way, with its activities covering four key areas: Software Engineering, Very Large Scale Integration, Man-Machine Interfaces and Intelligent Knowledge Based Systems.

Other programmes are in operation elsewhere in continental Europe and the EEC has established its own programme of international collaboration known as Esprit.

Whatever the outcome of these initiatives, their most striking feature is simply their existence. Without the Japanese fifth generation programme, it is doubtful whether any of them would ever have been conceived.

Only time will tell whether the Japanese project is successful, and the details will doubtless change as it proceeds, but it has already achieved the considerable feat of turning the spotlight on little-known but important work in the West. It is hard to disagree with the comment of Bob Muller of SPL International that:

> At the very least, Japan has set the world computing targets for the rest of the decade and beyond.

References

Alvey Committee (1982). A programme for advanced information technology: The report of the Alvey Committee. HMSO.

Bramer, M. A. (1982). A survey and critical review of Expert Systems research. In Michie, D. (ed). Introductory readings in Expert Systems. Gordon and Breach.

Clocksin, W. F. and Mellish, C. S. (1981). Programming in Prolog. Springer-Verlag.

JIPDEC (1981). Preliminary report on study and research on fifth generation computers 1979–1980. Japan Information Processing Development Center.

Shortliffe, E. H. (1976). Computer-based medical consultations: MYCIN. New York, American Elsevier/North Holland.

Bibliography
A | *Principal Documents*

Abstract Numbers: A1-A13

A1 **Alvey VLSI and CAD Strategy**
Alvey Directorate (December 1983)

This paper outlines the proposed strategy to be followed in the VLSI and CAD areas of the UK government's Alvey Programme on advanced information technology. It represents the first policy statement by the Alvey Directorate on the VLSI and CAD Strategy and will be updated as the programme progresses in the light of comments received. The overall objective in this sector of the Alvey programme is to carry out that research necessary to establish by the late 1980s internationally competitive VLSI processes within the UK. To achieve this objective it is intended to carry out co-ordinated research on a range of leading-edge technology options together wuth necessary advanced CAD tools.

Keywords: **Alvey programme, CAD, United Kingdom, VLSI**

A2 **The Fifth Generation — Dawn of the Second Computer Age**
SPL International

The proceedings of an international conference held in London in July 1982 by SPL International. The proceedings comprise twenty papers plus printed copies of the transparencies used by each speaker, making a total length of over 500 pages.

A description of each of the principal items in the proceedings is included separately below.

(See also B11, B24, B40; C3, C5, C9; D17, D38, D41, D45, D48; E20, E32; F43, F57, F59; G3)

Keywords: **collection of papers**

A3 **Fifth Generation World Conference, 1983**
SPL International

The proceedings of an international conference held in London in September 1983 by SPL International.

The proceedings comprise nine complete papers, together with notes from other presentations, plus printed copies of the transparencies used by each of the conference speakers. A copy of (A6) and its supplement is also included.

A description of each of the principal items in the proceedings is included separately below.

(See also A6; B1, B12, B13, B17, B31; F55, F61)

Keywords: **collection of papers**

A4 **Intelligent Knowledge Based Systems: A Programme for Action in the UK**
Alvey Directorate (August 1983)

A three-volume report produced by an intensive study commissioned for the UK government's Alvey Programme. The study drew on both academic and industrial expertise and has been taken as the statement of the Alvey Programme's strategy for the Intelligent Knowledge Based Systems area.

Volume 1 is the main report and sets out the strategy for the proposed programme. Volume 2 comprises a number of reports on individual subtopics.

The remaining volume contains Annexes to Volume 1.

Keywords: **Alvey programme, IKBS, United Kingdom**

A5 **Outline of Research and Development Plans for Fifth Generation Computer Systems**
ICOT — Institute for New Generation Computer Technology, Tokyo (May 1982)

Summary of the Japanese research and development plans for the Fifth Generation computer project which covers the social and technical background to the project, the research topics and targets, the overall research and development plans and underlying philosophies concerning international cooperation.

(See also A6; B25)

Keywords: **FGCS project, international relations, Japan, social context**

A6 **Outline of Research and Development Plans for Fifth Generation Computer Systems (Second Edition)**
ICOT — Institute for New Generation Computer Technology, Tokyo (April 1983, with supplement dated September 1983)

This paper is a second edition of (A5), with the same title but substantially different content. It briefly sets out the reasoning behind the development of the fifth generation computer systems project, identifies the functions required by such systems — problem solving and inference, knowledge base, intelligent interface and intelligent programming — and the innovative hardware architecture and software necessary to achieve these functions, and describes the plans, both overall and in the initial stages, for achieving the project's goals.

The supplement is entitled 'Report of FGCS Project's Research Activities, 1982' and summarises the project's activities during 1982.

(See also A3, A5)

Keywords: **architecture, FGCS project, inference, intelligent interface, Japan, knowledge base , problem solving**

A7 **Proceedings of Research Area Review Meeting on Intelligent Knowledge-based Systems**
Science and Engineering Research Council (1982)

The proceedings of a two-day review meeting held in London in September 1982 to discuss the content of a proposed Specially Promoted Programme in Intelligent Knowledge Based Systems by the UK Science and Engineering Research Council.

The proceedings principally comprise reports of discussions in 'Syndicates' on: knowledge representation; inference; natural language; vision and object manipulation; man-machine interface; Expert Systems (1 and 2); machines and programming: research; machines and programming: common base, and in discussion groups on: research programme and projects: groupings; infrastructure requirements; training and education; interfaces to other areas.

Keywords: **education and training implications, expert system, IKBS, inference, knowledge representation, man-machine interface, natural language, United Kingdom, vision**

A8 **A Programme for Advanced Information Technology: the Report of the Alvey Committee**
Her Majesty's Stationery Office (1982)

This is the report of a committee set up by the United Kingdom Department of Industry to advise on the scope for a collaborative research programme in information technology and to make recommendations relating to a UK response to the Japanese Fifth Generation Computer Programme.

The report recommends a five-year programme to mobilise the UK's technical strengths in information technology through a Government-backed collaborative effort between industry, the academic sector, and other research organisations. The goal is a strong UK capability in the 'core'-enabling technologies, which is considered essential to Britain's future competitiveness in the world information technology market.

Keywords: **Alvey programme, United Kingdom**

A9 Research Reports in Japan: a Collection of Recent Research Reports related to the R & D of the Fifth Generation Computer Systems

Japan Information Processing Development Center (Fall 1981)

A collection of 28 reports, on the following topics: deriving a formal specification of a problem from its natural language description; deriving a program from its formal specification; aspects of logic programming (12 papers); machine architecture (9 papers); speech recognition (2 papers); knowledge representation languages; machine translation; network-oriented operating systems.

Keywords: collection of papers, FGCS project, Japan

A10 Moto-Oka, T.
Challenge for Knowledge Information Processing Systems (Preliminary Report on Fifth Generation Computer Systems)
In (A11), pp3-89

Originally published by the Japan Information Processing Development Center with the title 'Preliminary Report on Study and Research on Fifth Generation Computers 1979-1980' dated Fall 1981, but more popularly known as the JIPDEC report.

This report sets out the Japanese research and development plans for the coming decade towards Fifth Generation Computer Systems. It is set in the context of the social problems to be faced by Japan in the 1990s and the consequent functional requirements for the new generation of computer systems. Such systems will be knowledge information processing systems having problem-solving functions of a very high level. Intelligence will be greatly improved in such systems to approach that of human beings and, in comparison with conventional systems, the man/machine interface will become closer to the human system. 26 themes in research and development are identified, grouped under the headings of basic application systems, basic software systems, new advanced architecture, distributed function architecture, VLSI technology, systematisation technology and development supporting technology. For each theme, details of the specific tasks to be performed and the target specifications to be achieved are set out. The project will be divided into initial, intermediate and final stages and the respective research items will be interrelated and mutually adjusted at the beginning of each stage. The role of international cooperation in such a project is stressed although it is seen essentially as a national project.

Keywords: **architecture, distributed processing, FGCS project, international relations, Japan, knowledge engineering, man-machine interface, problem solving, social context, VLSI**

A11 Moto-Oka, T. (ed)
Fifth Generation Computer Systems: Proceedings of the International Conference on Fifth Generation Computer Systems, Tokyo, Japan, October 19-22, 1981
North Holland Publishing Co. (1982)

The proceedings of an international conference held in Tokyo on October 19–22 1981 by the Japan Information Processing Development Center (JIPDEC) at which the Japanese Fifth Generation Computer Programme was announced. The proceedings comprise 18 papers under the section headings 'Keynote speech', 'Overview report', 'Knowledge information processing research plan', 'Architecture research plan', 'Invited lectures — knowledge information processing', and 'Invited lectures — architecture'. A description of each of the papers is included separately below.

(See also A10; B15, B20; C13, C17; D2, D3, D5, D33, D39, D41, D51; E37; F11, F25, F27, F35, F60)

Keywords: **collection of papers, FGCS project, Japan**

A12   Scarrott, G. G. (ed.)
The Fifth Generation Computer Project: State of the Art Report
Pergamon Infotech Ltd. (1983)

This state of the art report is divided into three parts: invited papers, analysis, bibliography. The invited papers examine various aspects of the Fifth Generation Computer Project. The analysis section assesses the major advances of the Fifth Generation Computer Project and provides a balanced analysis of the state of the art in the fifth generation. It is constructed by the editor of the report to provide a balanced and comprehensive view of the latest developments in fifth generation computer technology and is supplemented by quotations from the invited papers and current literature written by leading authorities on the subject. The analysis begins by considering the role of information in human affairs and the present status of information engineering. It goes on to discuss the techniques, hazards and limitations of technological forecasting in general and the application of such techniques in the information engineering field. This provides the conceptual framework in which the JIPDEC proposals are finally assessed. The annotated bibliography is a specially selected compilation of published material on the subject of the fifth generation.

(See also B21, B33, B42; C1, C14, C18; D4, D21, D26, D40; E17; F7, F56)

Keywords: **bibliography, collection of papers, FGCS project, Japan, social context**

A13   Sparck Jones, K.
Intelligent Knowledge Based Systems: Papers for the Alvey Committee
University of Cambridge Computer Laboratory (June 1982)

These papers were prepared as input to the Alvey Committee (see (A8)), at the request of one of its members. The papers consider intelligent knowledge-based systems under five headings: need, definition, status, programme, and outcome.

A major 10-year research and development effort in intelligent knowledge-based systems at a cost of £67M is proposed, as the core of the Alvey

Programme. This is envisaged as having two phases: an introductory phase, to establish the basic infrastructure for the research and development effort, and a main phase for the programme proper, with milestones set for the medium- and long-term to allow proper formulation and evaluation of the academic programmes and suitable 'demonstrator' projects.

Keywords: **Alvey programme, IKBS, United Kingdom**

B | Overview and Background Papers

Abstract Numbers B1-B45

B1 The Alvey Programme: Outline Plans
In (A3)

This short paper describes the outline plans of the British Government's Alvey Programme, based on the Alvey Directorate's progress report issued in July 1983 and its draft programme for Intelligent Knowledge Based Systems issued in August 1983. It is divided into ten sections: funding, management and organisation, major demonstrator projects, the technology programme, co-operation and participation, infrastructure and communications, information dissemination, property rights and exploitation, evaluation criteria and the relationships to the EEC Esprit Programme. The 'technology programme' section looks in turn at each of the four 'Alvey areas' of VLSI, Software Engineering, Intelligent Knowledge Based Systems and Man Machine Interfaces.

Keywords: **Alvey programme, IKBS, man-machine interface, software engineering, United Kingdom, VLSI**

B2 Japan (Special Report) Part I
Computing Europe, July 29 1982 pp.13-20

This special report on Japan's computer industry concentrates on several government-inspired projects aimed at pushing Japanese technology to its limits. The report opens with consideration of the Fifth Generation project — how it is to be achieved and its implications for international co-operation. The role of Japan's Ministry of International Trade and Industry (MITI) is next discussed, in particular in steering semiconductor manufacturers into productivity and in the developing area of supercomputers.

Keywords: **FGCS project, international relations, Japan, MITI, supercomputers**

B3 Japan (Special Report) Part II
Computing Europe, August 5 1982 pp.16-21

Second part of a special report on Japan's computer industry which focuses on the reaction to Japanese technology. Topics discussed include: the FBI allegations against Hitachi; anti-Japanese sentiment that has surfaced in the US following accusations of theft of US technology secrets by the Japanese; the Western belief in Japan's lack of innovativeness and contrary evidence regarding areas of creative excellence and recent moves by Japan's Ministry of International Trade and Industry (MITI) to stimulate software development; the booming high technology trade between Japan and Australia; Japanese plans for making use of dataflow machine architectures, parallel processing, relational databases and artificial intelligence techniques.

Keywords: **artificial intelligence, Australia, dataflow, international relations, Japan, MITI, parallel processing, relational database, United States of America**

B4 **MCC Moves out of the Idea Stage**
Science, Vol. 220 No. 4603 pp.1256–1257 (June 17 1983)

Microelectronics and Computer Technology Corporation (MCC) was created explicitly as a United States response to the Japanese Fifth Generation Computer Systems project. This paper examines its origins and progress to date. The joint venture corporation funds research and contributes researchers in four major areas: computer-aided design and manufacturing (CAD/CAM), computer architecture, software technology and packaging. Issues considered include MCC's encounter with antitrust policy, how the Austin site was selected, possible future involvement of European and Japanese companies, the problem of protecting research initiatives, the relationship with universities and federal government, and recruitment of staff.

Keywords: **architecture, CAD, CAM, international relations, MCC, software engineering, United States of America**

B5 **Tomorrow's Computers**
IEEE Spectrum, Vol. 20, No. 11, pp.34–120 (November 1983)

Special issue of the journal which comprises an introduction by Edward A. Torrero, 33 separate items on various aspects of fifth generation computing, and an annotated list of additional sources of information. The introductory article briefly describes the Japanese project proposals and international reactions to them. The remaining items are grouped into three sections: the quest, the challenges, and the outlook. 'The quest' comprises items by Robert E. Kahn, Robert J. Douglass, Trudy E. Bell, Tohru Moto-oka, Mark A. Fischetti, George A. Keyworth II, Robert S. Cooper, Brian W. Oakley, Horst Nasko, Erich Bloch, James D. Meindl and Richard B. Farr. It presents a glimpse of the technology of tomorrow and describes how most of the advanced nations of the world — Japan, the United States, Great Britain, and other European countries — are organising to develop the next generation of computers. This section includes an item on the European Strategic Programme on Research in Information Technology (ESPRIT) and two tables which present profiles of next generation computing research being carried out in universities and companies in the United States and various European nations (Belgium, Denmark, France, Great Britain, Greece, Ireland, Italy, the Netherlands, West Germany). Research concerned with both hardware and software aspects is included in the tables and an indication of the source of funding of each project is given. 'The challenges' comprises items by Paul Wallich, Robert J. Douglass, Fred Guterl, Edward A. Feigenbaum, Frederick Hayes-Roth, David L. Waltz, Raj Reddy, Victor Zue, Takeo Kanade, Raymond Yeh, Beau Sheil, A. L. Davis, Stephen Trimberger, and R. F. W. Pease. It shows how difficult a task researchers face and concludes with an assessment of the likely technological fallout. Topics discussed include computer networks, expert systems, natural language processing, speech recognition, computer vision, software engineering, computer architecture, and VLSI. The final item in this section is a report of a discussion of national priorities for next generation computer research by a group of U.S. scientists and engineers that took place via electronic mail on the CSnet, a National Science Foundation-sponsored network connecting computer science researchers. 'The outlook' comprises items by John Naisbitt, Fred Guterl, Erich Bloch, and Rob King. It presents a sociotechnological assessment of what the future holds and how institutions and individuals alike can prepare for the future. The item by Fred Guterl is a report of a round-table discussion of the projected impacts of next generation computing by eight engineers and social scientists from industry, government, and academia.

Keywords: **architecture, artificial intelligence, Belgium, bibliography, collection of papers, Denmark, Esprit programme, expert system, FGCS project, France, Greece, Ireland, Italy, Japan, knowledge representation, natural language, Netherlands, networks, social implications, software engineering, supercomputers, United Kingdom, United States of America, vision, VLSI, voice recognition, West Germany**

B6   Balzer, R., Cheatham, T. E. Jr. and Green, C.
Software Technology in the 1990s: Using a New Paradigm
Computer, Vol. 16, No. 11, pp.39–45 (November 1983)

In the past computers were more expensive than people; now, people are more expensive. Thus, there is a clear need to investigate a software paradigm based on automation, which augments the effectiveness of the costly and limited supply of people producing and maintaining software. The technology needed to support such a paradigm does not yet exist. The characteristics of an automation-based software paradigm are specified through an examination of the objectives it must meet. Such a paradigm is characterised as having formal specifications created and maintained by end users, with revised specifications becoming prototypes of the desired system thus ensuring responsiveness to user needs. The specification is reimplemented after each revision and, thus, the implementation process must be fast, reliable, and inexpensive. Fully automatic implementation via a computer is seen as non-feasible. A partially automated process is described, together with the way in which this automated implementation support might fit within a more general automated assistant structure. The article concludes by examining the sociological benefits of adopting such a new paradigm.

Keywords: **software engineering**

B7   Bird, J.
Fifth Generation (Special Report)
Computing Europe, November 4 1982 pp.19–28

Special report on Fifth Generation computing which looks at current projects and products and the pitfalls that are emerging. Topics considered include the following: the reasons for developing Expert Systems; the kind of products that should be produced initially, constraints on building systems and some unfounded criticisms; the role of logic programming in Expert System development; the computer education and training implications of the new computer systems; the possible Fifth Generation computers being developed in the UK at Imperial College, London (a graph reduction machine) and at Manchester University (a dataflow machine); the role of co-operation both among UK companies and across national boundaries; The work of Inmos and Isis Systems — the first UK Expert System company to receive government funding; ICL's knowledge-engineering project; the importance of VLSI technology for UK developments.

Keywords: **dataflow, education and training implications, expert system, graph reduction, ICL, Imperial College, Inmos, Isis Systems, knowledge engineering, logic programming, Manchester University, United Kingdom, VLSI**

B8   Boley, H.
Artificial Intelligence Languages and Machines
Techniques et Science Informatiques, Vol. 2 No. 3 pp.145–166 (May-June 1983)

The needs of Artificial Intelligence (AI) researchers are different from those of classical numerical programming because they aim to implement software for the non-algorithmic part of intellectual activities. As early as 1956 with IPL 1 and 1958 with Lisp 1, new languages began to appear. In the following years, formal calculus, scheme building, problem solving, logical reasoning, database interrogation, natural language analysis and synthesis, knowledge base management, etc. presented new needs which progressively have led to new concepts, processes and programming styles.These are embodied in high-level languages along with dedicated machines. In this paper, the author looks at the main concepts and then proposes a taxonomy of these languages and machines. He also shows their potential outside the realm of AI, and discusses their future impact on fifth generation projects.

Keywords: **artificial intelligence**

B9 Brandin, D. H.
The Challenge of the Fifth Generation
Communications of the ACM, Vol. 25, No. 8, pp.509–510 (August 1982)

This paper describes the characteristics of the computer systems that are to be developed under the Japanese Fifth Generation computer project, together with the social and technical goals of the project, and considers the implications for the US computer industry. It is argued that the Japanese goals are largely achievable and that the West must respond with comparable investments in research and facilities, because of the benefits inherent in a technological lead. The particular role of the ACM in such developments is discussed finally.

Keywords: **ACM, FGCS project, Japan**

B10 Connolly, R.
U.S. R & D Consortium Takes Shape
Electronics, Vol. 55, No. 5, pp. 97–99 (March 10 1982)

This paper outlines a plan by the chairman of Control Data Corporation to counter the threat posed by Japan's Fifth Generation computer project to the US semiconductor and computer industries. The plan involves setting up a multicompany research and development consortium to be known as MCE Inc.

Keywords: **MCE Inc., United States of America**

B11 d'Agapeyeff, A.
An Introduction to the Fifth Generation
In (A2)

An introductory paper from the conference chairman, providing a background to the Fifth Generation proposals. Topics covered include 'innovation versus stability', 'technical trends and the Japanese synthesis' and 'commercial implications'. Printed copies of the transparencies used by the author in his conference presentation are appended to the paper.

Keywords: **FGCS project, Japan**

B12 d'Agapeyeff, A.
Preparing for Fifth Generation Computing
In (A3)

In this paper the author considers the practicalities of introducing knowledge-based systems into a company already using conventional computing methods, in particular the technical and political difficulties of introducing radical change. The discussion is in the form of an allegory of an imaginary computer-using company with a resident Believer following a fifth generation Vision embarking on the perilous waters of a First Project.

Keywords: **business application, knowledge based system**

B13 Feigenbaum, E. A.
Worldwide Fifth Generation Computing: Developments, Issues and Concerns
In (A3)

Paper prepared for the SPL 1983 Fifth Generation World conference but not presented. Feigenbaum first sketches the scene in the Japanese fifth generation project on the basis of his visit there in July and August 1983, then looks at emerging American activities in areas that are called fifth generation in Japan and finally offers an essentially pessimistic view of the Alvey Initiative in the United Kingdom.

Keywords: **Alvey programme, FGCS project, Japan, United Kingdom, United States of America**

B14 Feigenbaum, E. A. and McCorduck, P.
The Fifth Generation: Artificial Intelligence and Japan's Computer Challenge to the World
Addison-Wesley (1983)

A best-selling book by Edward Feigenbaum, the head of the Stanford University Heuristic Programming Project, and Pamela McCorduck, a professional science writer. The book is written at a 'popular' level, aiming at explaining the development and the underlying aims of the Japanese Fifth Generation project in a non-technical fashion. The United States government and computer industry are urged to act vigorously to respond to the Japanese challenge.

The book is controversial in its assessment of industry, government, education and the status of Artificial Intelligence in Japan, the United States, Britain, and Continental Europe.

Keywords: **FGCS project, Japan, United Kingdom, United States of America, Western Europe**

B15 Fuchi, K.
Aiming for Knowledge Information Processing Systems
In (A11), pp.107-120

In constructing images for Fifth Generation computers, the author argues that one of the necessary conditions will be a survey of basic theoretical research, including research into the mathematical theory of programming, artificial intelligence and pattern information processing. The new generation of computers will not simply be an extension of existing ones but a definite leap forward, as a means towards coming to terms with the dissatisfaction felt over the basic structure of existing systems. A number of problems with existing systems are discussed, together with the kind of research that they necessitate. A major problem is that today's technology is far from the ideal of being truly 'handy' for users and the future research themes identified are natural languages including Japanese, the representation of knowledge and inference, software engineering and new computer architectures.

Keywords: **architecture, artificial intelligence, inference, knowledge engineering, knowledge representation, natural language, software engineering, theory of programming**

B16 Fuchi, K.
The Direction the FGCS Project Will Take
New Generation Computing, Vol. 1, No. 1, pp.3-9 (1983)

This paper briefly outlines the purpose and direction of the Japanese Fifth Generation Computer Systems project run at the ICOT research centre. The project is aimed at accomplishing basic research for the development of new technology rather than producing commercially-oriented products. Justifications are offered for the decisions to aim for innovation rather than the improvement of current computer technology, to emphasise logic programming rather than functional programming, to emphasise logic in the approach to knowledge engineering and to include hardware research in the project. It is argued that the direction being taken is the best possible way to meet the needs of Japanese future society.

Keywords: **FGCS project, Japan, knowledge engineering, logic programming, social context**

B17 Gannon, T. F.
Background Paper on the Micro Electronics and Computer Technology Corporation (MCC)
In (A3)

MCC will start operations with four technology programmes, namely the development of software productivity tools, an integrated system of computer-aided design and manufacturing tools, a program to investigate advanced computer architecture, and systems and chip-level packaging technology. This paper presents the case for work in each of these areas, what each programme will produce, and details of costings and timescales.

Keywords: **architecture, CAD, CAM, MCC, United States of America**

B18 Hudson, K.
How We Can Weather the Brainstorm from Japan
Computing, Vol. 11, No. 36, pp.16-17 (September 8 1983)

The author discusses the areas of research which are expected to lead to fifth generation computers. The Japanese fifth generation project is compared with the Alvey initiative in Britain. The author feels that the Alvey Committee's decision to concentrate on software development is a mistake. He believes that the Japanese plan to work on hardware and software simultaneously is a better idea. Even so, the author feels that the

Americans are most likely to reap benefits from artificial intelligence research because of their extensive work in the brain sciences. The author believes that no significant breakthrough will be achieved in expert systems until more is known about the human brain.

Keywords: **Alvey programme, expert system, FGCS project, human knowledge processing, Japan, United Kingdom, United States of America**

B19 Johnson, J.
Can They Do It?
Datamation, Vol. 29, No. 7, pp.161-170 (July 1983)

Report of interviews with Peter Gregory, Marty Goetz, Ted Withington and Jonathan Allen, four people from different segments of the United States computer industry, concerning their views on the Japanese fifth generation project. Each person was asked about his perceptions of the project's motives and objectives, the likelihood of success, how Japan's research efforts compare with those of the United States and the speed at which progress is likely to be made in each country, the strengths and weaknesses of the United States in the face of the Japanese challenge, how the United States computer industry should view the Japanese proposals and what action should be taken to protect its position, and the effects of the Japanese proposals on their own work and that of their employing companies.

Keywords: **FGCS project, Japan, United States of America**

B20 Karatsu, H.
What Is Required of the Fifth Generation Computer: Social Needs and Its Impacts
In (A11), pp.93-106

The author identifies the bottlenecks that must be overcome in Japan if his model of a desirable society is to be realised. These bottlenecks include areas of low productivity, lack of internationalisation, limited natural resources, an aged and highly educated population and great dependence on information.

He goes on to outline the requirements for Fifth Generation computers if they are to come to terms with these problems in relation to the man/machine interface, speed of information processing, artificial intelligence applications, database design and use, and communication networks. The article concludes by considering the impact of the new machines on the individual, the social system, industry and international relations.

Keywords: **artificial intelligence, database, international relations, Japan, man-machine interface, networks, social context**

B21 Kawatani, Y.
Japan: Road to the Future
In (A12) pp.63-73

This paper describes the work which has been going on in Japan to lay the foundation for the future progress of Japan's high technology industries. Developments at Fujitsu Limited are assessed, with particular reference to semiconductor technology and Fujitsu's High Electron Mobility Transistor. A revolutionary improvement in Japanese word processing and voice recognition technology has been taking place and this is described. The importance of introducing industrial robots and computer-aided manufacturing for enhancing reliability and productivity of computer-related equipment is stressed, together with the significant role of employees' dedication to quality control in achieving high product reliability.

Keywords: **CAM, Fujitsu, Japan, robots, social context, voice recognition**

B22 Kim, K. H.
A Look at Japan's Development of Software Engineering Technology
Computer, Vol. 16, No. 5, pp.26-37 (May 1983)

This paper is based predominantly on observations made and material obtained during a nine-day visit to ten Japanese organisations engaged in software engineering R & D: three government-operated/supervised organisations (Information-Technology Promotion Agency, Yokosuka Electrical Communication Laboratory, Electro-Technical Laboratory), four industrial laboratories (Fujitsu Labs, Ltd, Hitachi Software Engineering Co. Ltd, Nippon Electric Company, Toshiba Fuchu Works) and three universities (Keio University, University of Tokyo and Tokyo Institute of Technology). In addition to describing the work at each of these institutions, information is provided on Japan's national R & D projects with particular reference to the projects on Very High Speed Scientific Computing Systems (supercomputers) 1981-1988 and Fifth-Generation Computer Systems, 1982-1991. The author concludes by identifying three major trends in the development of Japan's software engineering technology: 1) a focus on testing and integrating various concepts/approaches originated in the United States and Europe; 2) an emphasis on producing tangible end-user products; 3) particular advance in real-time software technology and basic artificial intelligence technology, including pattern recognition. Although the United States does not appear to have lost its technological edge in software engineering, it is suggested that the utilisation of these technologies may be a different matter.

Keywords: **artificial intelligence, ETL, FGCS project, Fujitsu, Hitachi, Information-Technology Promotion Agency, Japan, Keio University, Nippon Electric Company, software engineering, Supercomputers** project, Tokyo Institute of Technology, Tokyo University, Toshiba Fuchu Works, Yokosuka Electrical Communication Laboratory

B23 Kowalski, R.
The Fifth Generation Project in Japan
British Computer Society Specialist Group on Expert Systems Newsletter, No. 8, pp.11–14 (May 1983)

Survey of activities related to the fifth generation computer project based on a British Council sponsored seventeen day visit to Japan at the end of 1982. Details are given of the work being carried out at ICOT (the Institute for New Generation Technology), NTT (Nippon Telephone and Telecommunications) at Yokosuka and Masashino, ETL (the Electrotechnical laboratory at Tskuba), the universities of Tokyo, Tohoku and Kyoto, Tokyo Institute of Technology and NEC in Kawasaki. The paper concludes by identifying major research themes, in particular the preference for PROLOG in contrast to LISP for the main artificial intelligence application areas of natural language processing and expert systems, and the greater advances in the former application area as compared with the latter, within the more general context of a low priority to applications as against logic machines and their supporting programming environments.

Keywords: **artificial intelligence, ETL, expert system, FGCS project, ICOT, Japan, Kyoto University, logic machine, natural language, NEC, NTT, Prolog, Tohoku University, Tokyo Institute of Technology, Tokyo University**

B24 Lehman, M. M.
The Role of Systems and Software Technology in the Fifth Generation
In (A2). Also available as Imperial College of Science and Technology, Department of Computing, Research Report

The Japanese Fifth Generation computer plan gives only scant attention to the problems that have plagued two decades of the production and maintenance of reliable, cost-effective and timely software; problems that must have similar consequences in FGCS unless appropriate steps are taken from the start. This paper briefly examines the significance and implications of these problems under the headings of requirements, evolution, complexity, understanding correctness, responsiveness and cost. It is argued that in seeking to pursue the objectives of the Japanese plan, or its UK equivalent, adequate effort must be invested in software technology, to support the development, secure exploitation and continued evolution of further computer generations. The aim must be to benefit mankind. Inadequate software technology support could instead cause systems such as those proposed to constitute a force for society's decay. Printed copies of the transparencies used by the author in his conference presentation are appended to the paper.

Keywords: **FGCS project, software engineering**

B25 Lemmons, P.
Japan and the Fifth Generation
Byte, Vol. 8, No. 11, pp.394–401 (November 1983)

This article comprises a reprint of (A5), together with an introduction which analyses the proposals. The introduction challenges those American critics who see Japan's plans as analogous to a military attack and presents the Japanese view of how the plans came about and their goals. The plans are seen as a means by which a resource poor country might maintain its status as an advanced nation and a major goal is to have computers that are easy to use and can handle natural language. It is concluded that Japan will compete with the United States and other Western nations in information processing technology over the next decade, but it seems unlikely that Japan will dominate the field on the basis of the current project, given that many United States projects are more heavily resourced.

Keywords: **FGCS project, Japan, social context, United States of America**

B26 Malik, R.
Behind the Fifth Generation
Computerworld, Vol. 16, No. 46A, pp.23–27 (November 17 1982)

The author discusses the current state of the Japanese computer industry and the likelihood of the fifth generation computer systems project being a success. The technological advances needed to develop the new computers are discussed.

Keywords: **FGCS project, Japan**

B27 Malik, R.
In the Maze of the Fifth Generation
Computing, Vol. 11, No. 9, pp.23–25 (March 3 1983)

Japanese plans for a fifth generation computer are surrounded by considerable mythology, which the author attempts to dispel. He argues that the fifth generation is not just another computer development project for making the technology run faster or more cleanly or more simply, but an attempt to create a completely new family of general purpose machines. The fundamental roles of the Prolog language and parallel hardware machine architecture are described and it is concluded that, if the project is successful, the Japanese are likely to push all other competitors into a tiny corner of the computer market.

Keywords: **FGCS project, Japan, parallel processing, Prolog**

B28 Manuel, T.
Japanese Map Computer Domination
Electronics, Vol. 54, No. 23, pp.83–84 (November 17 1981)
Reprinted in *Byte*, Vol. 7, No. 5, pp.140–144 (May 1982) under the title
Japan Maps Computer Domination

This paper summarises the characteristics of the Japanese Fifth Generation Computer System project which neatly integrates many of the innovative ideas from researchers in the US, Japan, and the rest of the world. The three basic functions of the new computers will be the ability to handle a wide range of general problem-solving targets through the 'intelligent-interface machine', the ability to learn, associate and infer through the problem-solving and inference system, and the ability to use stored information through the knowledge-base management system. The computers will come in all sizes interconnected with local and global networks and will employ new architectures and languages. The R & D themes identified for the project are listed, together with some of the target specifications.

Keywords: **architecture, FGCS project, inference, intelligent interface, Japan, knowledge base, machine learning, networks, problem solving, United States of America**

B29   Manuel, T.
West Wary of Japan's Computer Plan
Electronics, Vol. 54, No. 25, pp.102–104 (December 15 1981)

This paper presents some views of Western university and industry experts on the Japanese Fifth Generation Computer System project. The general opinion seems to be that, although the project is ambitious, it will be successful and that it is not a good idea to be too complacent.

Keywords: **FGCS project, Japan**

B30   McCorduck, P.
An Introduction to the Fifth Generation
Communications of the ACM, Vol. 26, No. 9, pp.629–630 (September 1983)

This brief introduction to the Japanese fifth generation computer systems project: 1) describes the three successive stages of the project's ten-year plan, concentrating in particular on the research and development plans for intelligent interfaces, through which fifth generation machines will be easier to use than existing ones, and 2) gives an account of the innovative administrative organisation of the project, whereby leading researchers from the consortium of eight firms and two national bodies have come together under one roof to work together. It concludes by considering the likely outcomes of the project and the reasons for its appeal. It is argued that fifth generation machines are not only possible but inevitable, and that the appeal of the project lies in the noble vision of the future suggested — a future where knowledge is the wealth of nations.

Keywords: **FGCS project, ICOT, intelligent interface, Japan**

B31 Michie, D.
The Turing Institute
In (A3)

A few years ago the author and a number of colleagues were involved in preparing a prospectus for a research institute for advanced computer science and engineering. More recently the Alvey recommendations have led to the establishment of around thirty Information Technology (IT) fellows at academic centres, forming what might be termed a dispersed national IT institute. The author believes this does not rule out the possibility of a single centre but that this should be funded predominantly by industry rather than government. This paper reproduces much of the original text with appropriate revisions, mostly in the finance section, and expands on steps the author has been taking towards implementing the prospectus, spurred on by the sense of urgency imparted by the Japanese Fifth Generation Project. The prospectus establishes the need for an institute and sets out its objectives, its programme of work including activities in the areas of computer architecture, automatic programming, expert systems and advanced robotics, its organisation and its finance. As a first step, a not-for-profit company has been established, Machine Intelligence Research Affiliates, as a revenue-earning workhorse which will carry on its back the advanced study programme — the Turing Institute itself. The company offers in-depth support to its industry members and carries out contract research activities.

Keywords: **architecture, automatic programming, expert system, MIRA, robotics, Turing Institute, United Kingdom**

B32 Mill, J.
Cracks Appear in the AI Gravy Train
Informatics, Vol. 4, No. 10, pp.23-24, 26 (October 1983)

The author investigates through conversations with leading members of the AI community, why the optimistically founded Alvey and Esprit funding programmes are no guarantee that Britain will win the AI race. Major problems are the 'brain-drain' among British artificial intelligence experts and the risk of research contracts going to organisations that have sprung up overnight to take advantage of the AI boom and which have no internal expertise. In addition the Government's decision to provide a maximum of 50% funding for research projects within the Alvey programme instead of the 90% originally recommended means that small innovative companies are unlikely to be involved. The current projects being funded under the Esprit programme are described, together with independent work at the Imperial Cancer Research Fund and Export Software International. Problems associated with pursuing theoretical rather than application based research and with the lack of awareness among the AI community of relevant psychological research are identified. Finally the importance of some of the tools developed by AI researchers for those in other application areas is noted.

Keywords: **Alvey programme, artificial intelligence, Esprit programme,**

Export Software International, Imperial Cancer Research Fund, United Kingdom

B33 Moto-Oka, T.
Overview and Introduction to the Fifth Generation
In (A12) pp.3–16

This overview and introduction to the Japanese fifth generation computer (FGCS) project describes the social demands that the new computers must meet, the technical background, the research and development targets of the project in relation to problem-solving and inference, knowledge base management and intelligent interface, the basic configuration of the FGCS and the application systems to be developed for machine translation, consultation, intelligent programming and intelligent VLSI-CAD. The paper concludes by stressing the need for international co-operation and suggesting the form this might take.

Keywords: **automatic programming, CAD, consultation system, FGCS project, inference, intelligent interface, international relations, Japan, knowledge base, machine translation, problem solving, social context, VLSI**

B34 Shapiro, E. Y.
The Fifth Generation Project: A Trip Report
Communications of the ACM, Vol. 26, No. 9, pp.637–641 (September 1983)

The author provides insights into the fifth generation project, based on a trip to the Institute for New Generation Computer Technology (ICOT) in Japan from October to November 1982. The particular focus of attention is the Japanese commitment to build the fifth generation system around the concepts of logic programming. The report traces the roots and rationale for this commitment and its possible implications.

Keywords: **FGCS project, ICOT, Japan, logic programming**

B35 Simons, G. L.
Towards Fifth-Generation Computers
NCC Publications (1983)

An introductory book which leads the reader through the different themes of the Fifth Generation project. As well as Artificial Intelligence and Expert Systems, it covers trends in integrated circuit technology, parallel architectures such as dataflow, logic programming, voice recognition and visual perception. The response to fifth generation plans is briefly surveyed, with focus on the Alvey Report. Some speculation is included about future possibilities.

Keywords: **Alvey programme, architecture, artificial intelligence, dataflow, expert system, integrated circuit technology, logic programming, United Kingdom, vision, voice recognition**

B36 Spennewyn, D.
Where Super Systems Go
Computing Europe, October 28 1982 pp.32-33

This article argues that, whereas in the past the term 'supercomputer' implied one which could perform a relatively large number of instructions per second, today the term might better be applied to computers which offer good facilities to the user or perform highly specialised data processing tasks requiring vast amounts of computation. Two obvious application areas for supercomputers in this latter sense of the term are pattern recognition and knowledge- based (expert) systems. The most widely used of today's supercomputers are the pipelined vector processors, the Cyber 205 and the Cray. This review of a recent State of the Art Report describes these two machines together with some of the less well developed supercomputers of the array processor design. It also discusses the benefits of distributed computing and the balance of application areas in relation to supercomputers.

Keywords: **array processor, Cray computer, Cyber 205 computer, distributed processing, expert system, pattern recognition, supercomputers, vector processor**

B37 Steier, R.
Cooperation is the Key: An Interview with B. R. Inman
Communications of the ACM, Vol. 26, No. 9, pp.642-645 (September 1983)

Report of an interview with retired Admiral B. R. Inman, the president of Microelectronics and Computer Technology Corp (MCC), one of the most innovative responses to Japan's fifth generation project. MCC is a consortium of thirteen companies formed to conduct long range research and development in advanced computers. Admiral Inman comments on the architectural approach being taken by MCC, its recruitment plans, its plans in relation to disseminating information, parallels with the Japanese fifth generation project, the relationship of MCC with the government, industry and the universities, plans for expansion and problems with Justice Department restrictions, the choice of the Austin site for MCC, the impact of MCC on technology transfer and the possible role of ACM in encouraging collaborative research.

Keywords: **ACM, FGCS project, MCC, technology transfer, United States of America**

B38 Sumner, F. H.
The Future of Information Technology: A Personal View
ASLIB Proceedings, Vol. 35, No. 1, pp.1-13 (January 1983)

Current research topics in information technology are examined and the author explains how he feels the technology will develop. He examines possible new developments in component technology, silicon-based and otherwise. A general overview of current ideas for future computer architecture is given. The potential of expert systems and other artificial intelligence ideas is examined. Finally a look at the Japanese fifth generation computer systems project gives a glimpse of things that may be to come.

Keywords: **architecture, artificial intelligence, expert system, FGCS project, information technology, Japan**

B39 Tanaka, K.
Artificial Intelligence and Computers
ICOT Journal, No. 1 (June 1983)

The author identifies the functional requirements for fifth generation computers that underly the Japanese Fifth Generation Computer Systems Project. He argues that to make viable systems that embody human activities such as thinking or creation calls imperatively for the synergy of software technology backed up by artificial intelligence or knowledge engineering, architecture technology (program, software and system architectures), and hardware configuration technology. This means the development of an integrated cognitive system which will certainly become a reality, though when depends on the priority given to it.

Keywords: **architecture, artificial intelligence, FGCS project, Japan, knowledge engineering, software engineering**

B40 Treleaven, P. C. and Lima, I. G.
Japan's Fifth-Generation Computer Systems
Computer, Vol. 15, No. 8, pp.79–88 (August 1982)
Also included in revised form in (A2) with the title **Japan's Fifth Generation Computer Systems Project**

This article describes the main aspects of the Japanese Fifth Generation Computer Systems project. These systems are intended to represent a unification of four currently separate areas of research, namely knowledge-based Expert Systems, very high-level programming languages, decentralised computing and VLSI technology. In addition to presenting an image of the Fifth Generation Computer System, the various subsidiary research projects involved are outlined and the likely impacts and effects of the research are discussed.

Printed copies of the transparencies used by the author in his conference presentation are appended to the version included in (A2).

Keywords: **distributed processing, expert system, FGCS project, Japan, VLSI**

B41 Waltz, D. L.
Artificial Intelligence
Scientific American, Vol. 247, No. 4, pp.101–105, 109–110, 112–114, 116, 119–122 (October 1982)

The author conveys an idea of the kind of results that can be achieved by programs in artificial intelligence. The themes examined include the use of heuristic principles to shorten searches, the relation between human and machine strategies, goal-directed planning, backward chaining, concept learning, constraint propagation and machine understanding of natural language. In examining these themes the author refers to a range of artificial intelligence programs in areas such as game playing, the blocks world, mathematical concept learning, the analysis of line drawings and language understanding. It is concluded that the most challenging task facing artificial intelligence is the modelling of common sense.

Keywords: **artificial intelligence, human knowledge processing, natural language**

B42 Warren, D. H. D.
A View of the Fifth Generation and its Impact
In (A12) pp.145–153

This paper gives a personal view of the Japanese Fifth Generation Computer Systems Project based on a range of sources of information. It begins by reviewing (B15) in order to explain the reasoning behind the Japanese approach to the fifth generation, with particular attention to how PROLOG came to be adopted as the kernel language. The first three years of the project plan are discussed with reference to progress reported from Japan, and British and American reactions. The paper concludes by reviewing the status of PROLOG in the United States, where work on PROLOG-based machines and applications is already under way at several centres.

Keywords: **FGCS project, Japan, logic machine, Prolog, United States of America**

B43 Weil, U.
Evaluating the Japanese Challenge
Datamation, Vol. 28, No. 13, pp.164–168 (December 1982)
Also in *Datamation*, Vol. 29, No. 1, pp.122–134 (January 1983)

This article asks whether the overseas successes in selected computer and telecommunications product areas of the Japanese is the beginning of a tidal wave of systems destined to subjugate the United States-based computer industry. Likely trends in Japan's position in the overseas hardware and software markets are considered.

Keywords: **Japan**

B44 Withington, F. G.
Winners and Losers in the Fifth Generation
Datamation, Vol.29, No. 12, pp.193–209 (December 1983)

Four areas of advanced computer technology are currently subject to particularly intense international competition: large-scale integrated circuits, disk drives (magnetic and optical), supercomputers and knowledge based systems. This article summarises forecasts of these technologies recently prepared for the United States federal government, predicts the degree of success that competitors in each region will have, and discusses their interrelationships in the future structure of the information industry. It is concluded that national governments will be among the winners in the fifth generation, with private companies and users possibly being among the losers. But restrictive national practices can only mean denying modern systems to a country's citizens, for the information industry of the fifth generation will be by its nature the most multinational in history.

Keywords: **international relations, knowledge based system, LSI, supercomputers**

B45 Yasaki, E. K.
Tokyo Looks to the '90s
Datamation, Vol. 28, No. 1, pp.110–115 (January 1982)

Japan is beginning a 10-year R & D effort on Fifth Generation Computer Systems. To delineate and defend the various aspects of the project, an international conference was held in Tokyo in October 1981. What the participants saw was a menu of research themes on computer architectures (particularly dataflow), programming languages (particularly PROLOG), knowledge engineering (Expert Systems with their knowledge bases and inference mechanisms) and facilities to handle speech I/O and image I/O — all implemented in VLSI. Natural language processing, language translation and database management were also mentioned. This paper presents some of the views of the conference attendees, some comments on the Japanese desire for participating governments to carry out parallel research and an account of some of the specific research projects to be carried out.

Keywords: **architecture, database, dataflow, expert system, FGCS project, image processing, Japan, knowledge engineering, machine translation, natural language, Prolog, speech output, speech understanding, VLSI**

C | *The Human-Computer Interface*

Abstract Numbers C1-C19

See also: F9

C1 Allen, J.
VLSI Applications: Speech Processing
In (A12) pp.41-48

If computing systems are to be broadly used by people without technical training, then it is essential that the user interface presented by the computing system bend to the needs of the human. Human speech is a natural basis to use for the man/machine interface but it is not ideal for communicating all kinds of information. Hence, increasing emphasis has been placed on the generation of speech from a computer and the recognition of human speech by a computer. In addition speech synthesis and recognition provide ideal tasks for the emerging VLSI technology. This paper examines constraints on speech synthesis and recognition including techniques needed to satisfy various system requirements; the nature of the human vocal apparatus; the need to utilise linguistic structures in speech recognition systems with input, and the need for a viable technology for the implementation of the various synthesis systems and recognition tasks. Different types of speech synthesis systems, ranging from simple systems which play back stored wave forms to comprehensive techniques for the conversion of unrestricted English text to speech, are examined in light of these constraints. Problems associated with research in the much more difficult area of speech recognition are finally examined.

Keywords: **speech output, VLSI, voice recognition**

C2 Ballard, D. H., Hinton, G. E. and Sejnowski, T. J.
Parallel Visual Computation
Nature, Vol. 306, No. 5938, pp.21-26 (November 3 1983)

The functional abilities and parallel architecture of the human visual system are a rich source of ideas about visual processing. The authors argue that any visual task that can be performed quickly and effortlessly is likely to have a computational solution using a parallel algorithm. This review looks at several recently developed parallel algorithms that exploit information implicit in an image to compute intrinsic properties of surfaces, such as surface orientation, reflectance and depth. The relevance of such algorithms to visual processing in the cerebral cortex of primates is discussed and future directions for research are considered.

Keywords: **human knowledge processing, parallel processing, vision**

C3   Bott, M. F.
A Future for Machine Translation
In (A2)

This paper begins by identifying the major reasons for the failure of early machine translation systems. It goes on to examine recent developments that make the development of such systems economically viable (in particular the increased use of word processors, which means that many documents' translations are available in machine-readable form) and developments in computational linguistics. The characteristics and performance of some currently available systems are described, together with those likely to be associated with future systems for some time to come. It is argued that it should now be technically feasible to produce translation systems working in tandem with word processing systems, which would produce economically attractive gains in the productivity of human translators. Such systems would only be capable of operating over a limited range of subject matter, but Expert Systems technology offers attractive possibilities for tailoring translation systems to new areas of discourse.

Keywords: **expert system, machine translation**

C4   Bruckert, E., Minow, M. and Tetschner, W.
Three-Tiered Software and VLSI Aid Developmental System to Read Text Aloud
Electronics, Vol. 55, No. 8, pp.133–138 (April 21 1983)

This article outlines an effort at Digital Equipment Corp, Massachusetts, to allow speech output from any text data base. Digital have adapted the three-tiered software approach developed by Dennis Klatt of the Massachusetts Institute of Technology to develop a two-processor board holding only a commercial microprocessor — Motorola's 6800 — and a digital signal-processing chip package — Texas Instruments TM32010. All software is written in the high-level language C and is transportable. The first of the three processing levels needed to turn text into speech generates unambiguous digital representations of speech sounds from 8-bit ASCII text, the second accepts these representations as input and from them calculates sets of acoustic parameters and the third synthesises the voice output under the control of these parameters. The system can be accessed through RS-232-C ports from either a local terminal or a host computer. The speaking rate is variable, from 120 to 130 words per minute. A choice of several voices and a loudspeaker telephone or analog output are provided.

Keywords: **Digital Equipment Corporation, speech output, VLSI**

C5 Duff, M. J. B.
Parallel Architecture and Vision
In (A2)

In designing a computer or processor a balance has to be struck between the extremes of an efficient but completely specialised system for a particular task and a flexible system giving average performance over a range of tasks. A middle course is now emerging in which the processor architecture is broadly structured to perform well for a certain class of operations but with deteriorated performance outside this class. In this area of image processing, CLIP4 is an example of such an architecture.

The Cellular Logic Image processor is typical of systems which approach the problem of parallelism by using an array of interconnected processors, ideally with a one-to-one relationship with the array of pixels in the picture to be processed. It was constructed at University College, London and is currently being applied to a variety of image analysis tasks. The team at University College is now investigating more sophisticated PE designs with a view to taking advantage of the opportunity to pack more gates on to an integrated circuit for little additional cost. Printed copies of the transparencies used by the author in his conference presentation are appended to the paper.

Keywords: **CLIP4, image processing, parallel processing, University College, vision**

C6 Dusek, L., Schalk, T. B. and McMahan, M.
Voice Recognition Joins Speech on Programmable Board
Electronics, Vol. 55, No. 8, pp.128–132 (April 21 1983)

This article describes a voice recognition and speech-synthesis board aimed at personal computers, the Texas Instruments SBSP3001 built around the high-speed TMS320 single-chip microcomputer. The 3001 uses the 320 to digitise and compress speech using a technique termed linear predictive coding and stores it in on-board or system memory in a compact form. Later, the compressed speech can be synthesised and played back.

Keywords: **personal computer, speech output, voice recognition**

C7 Erman, L. D., Hayes-Roth, F., Lesser, V. R. and Raj Reddy, D.
The Hearsay II Speech Understanding System: Integrating Knowledge to Resolve Uncertainty
Computing Surveys, Vol 12, No. 2, pp.213–253 (June 1980)

The problem of understanding connected continuous speech is discussed, together with the four systems developed during a five-year project funded by DARPA. The paper concentrates principally on the structure of the Hearsay II system which is also one of the most interesting from an Artificial Intelligence viewpoint. The presentation is aimed at a general computer science audience with no knowledge of the speech understanding or Artificial Intelligence fields.

Keywords: **Hearsay II, speech understanding**

C8   Evanczuk, S. and Manuel, T.
Practical Systems Use Natural Languages and Store Human Expertise
Electronics, Vol. 56, No. 24, pp.139–145 (December 1 1983)

This is the second part of a two-part special report on the commercial status of artificial intelligence which deals with software aspects, in particular natural-language interfaces and knowledge-based systems. A range of topics is covered including: 1) the technique of state space search as a means of making machines 'intelligent'; 2) the use of rules of grammar and augmented-transition-network parsers as means of addressing the problem of ambiguity in natural language; 3) developments in systems to access data bases, including techniques to personalise data bases and the use of menus to ensure user queries are accepted; 4) developments in language translation systems; 5) developments in knowledge based systems, such as Dendral, and in generic expert system development tools; 6) the view of artificial intelligence as being the field of symbolic computation and the role of functional programming languages such as Lisp; 7) techniques for constructing a correspondence between some sets of attributes and some stored data, such as association lists and frames; 8) object-oriented programming, particularly the language Smalltalk, and its relationship to Lisp. A note on programming in logic, in particular the use of Prolog, is incorporated in the article.

Keywords: **artificial intelligence, database, Dendral expert system, expert system shell, France, functional programming, knowledge-based system, Lisp, machine translation, natural language, object oriented programming, Prolog, search, Smalltalk**

C9   Harris, L.
The Four Obstacles to End User Computer Access
In (A2)

This paper argues that user-friendliness is not enough to ensure that computer systems will be used by non-technical end users. It addresses four major problems that must be overcome if any software product is to become a useful tool for large numbers of true end users: 1) the language problem; 2) the problems caused by the user's conceptual viewpoint; 3) the navigation problem; 4) the need to interface to multiple software systems. In discussing these problems the author draws a major contrast between a natural language approach to developing user-friendly systems, which involves determining what the end user wishes to accomplish and then providing the necessary technical support mechanism to allow the user to activate the process in a natural way, and the classical approach, which involves thinking first of the technical capabilities that can be provided and then of the most palatable command sequence to invoke it which the user must learn. Printed copies of the transparencies used by the author in his conference presentation are appended to the paper.

Keywords: **natural language, user-friendly system**

C10   Hendrix, G. and Sacerdoti, E.
Natural-Language Processing: The Field in Perspective
Byte, Vol. 6, No. 9, pp.304–352 (September 1981)

This paper offers an overview of the potential applications, experimental systems, existing techniques, research problems and future prospects in the rapidly evolving field of natural-language processing. Major issues in the field are addressed by focusing on several representative systems, in particular 1) LADDER, a system for translating English queries into a form which the machine understands; 2) SHRDLU, a system for dealing with dynamic microworlds of blocks; 3) SAM, a system for supporting reasoning tasks in relation to ordinary human situations using knowledge incorporated into formal constructs called scripts; 4) TDUS (Task-Oriented Dialogue Understanding System), a system for communicating with a human apprentice about repair operations on electromechanical equipment in which the information is recorded in data structures called procedural networks.

Keywords: **Ladder, natural language, SAM, SHRDLU, TDUS**

C11   Jain, R. and Haynes, S.
Imprecision in Computer Vision
Computer, Vol. 15, No. 8, pp.39–48 (August 1982)

In this paper the use of fuzzy set theory to solve certain problems in computer vision systems is shown, with the author's own dynamic scene system, Vili, used for the purposes of illustration.

The authors believe that computer vision systems should make use of approximate reasoning and both domain-dependent and domain-independent knowledge. It is claimed that exploiting uncertainty in data and using a model that parallels human reasoning processes can improve images in computer vision systems.

Keywords: **fuzzy sets, plausible reasoning, Vili, vision**

C12   Kaplan, S. J. and Ferris, D.
Natural Language in the DP World
Datamation, Vol. 28, No. 9, pp.114–120 (October 1982)

A lot of research has taken place over the last decade in the area of natural language processing and commercial products are beginning to become available. A major reason for this is that many believe that natural languages are the ultimate programming languages. Unfortunately, there is much evidence that this is not the case. This article examines some of the obstacles to the wider use of natural language database query systems and considers some of the common misconceptions about them. A main drawback is that natural language, entered by keyboard, is simply not an effective communication tool for many existing interactive applications. Natural language is often inefficient and ambiguous. Overcoming ambiguity requires that additional knowledge about the world with which queries are concerned be incorporated into the computer system. This creates a problem of portability across systems and for different query domains. Other problems include the need for systems to be able to deal with queries containing new words, to offer summary replies where appropriate and to overcome user scepticism. Misconceptions about such systems in contrast to artificial language systems include their high use of cpu time and their unreliability.

Keywords: **artificial intelligence, database query system, natural language**

C13   McCormick, B. H., Kent, E. and Dyer, C. R.
A Cognitive Architecture for Computer Vision
In (A11), pp.245-264

The authors describe a design for a visual analyser for real-time scene analysis of dynamic imagery, based on a cognitive architecture derived from concepts of the central nervous system. Output of the visual analyser will be high-level real-time perceptual data suitable either for the guidance of robot action or for the symbolic description of the visual environment. This output resides in a distributed relational database modelling the hypercolumnar architecture of the visual cortex. It is conjectured that the principles of cognitive computation described can be generalised to other sensory and effector information-processing tasks.

Keywords: **architecture, human knowledge processing, relational database, vision**

C14   Moto-Oka, T.
The Intelligent Interface System
In (A12), pp.101-114

The Japanese plan for research into intelligent man/machine interfaces is centred on three activities: natural language processing, speech processing and graph-image processing. In this paper primitive techniques to be developed during research into these areas are examined along with the role of natural language processing and speech processing in basic application systems. The graphic and image-processing functions described in the paper are implemented as system components of the intelligent interface: they can be divided into hardware and software components. The research and development plan of the Japanese Fifth Generation Computer System Project for intelligent interface systems is set out in an appendix.

Keywords: **FGCS project, image processing, intelligent interface, Japan, natural language, speech understanding**

C15 Neff, R.
Japanese Welcome Voice Recognition
Electronics, Vol. 55, No. 3, pp.97–98 (February 10 1982)

Japanese industry is moving firmly towards the adoption of voice recognition technology. So far the most popular uses, as in the US, are for distribution and inspection but a handful of firms is moving toward applications that are even newer. Examples are given of firms using the technology for part ordering, design, putting instructions on magnetic tapes that run machine tools and quality control. Applications in banks and for keyboard data entry are also discussed.

Keywords: **business application, Japan, voice recognition**

C16 Neff, R.
Machine Translation Regains its Voice
Electronics, Vol. 56, No. 4, pp.82–83 (February 24 1983)

This article describes developments in the field of machine translation, a technique once given up for dead but which has recently attracted much interest. Attention is focused in particular on 1) the Eurotra project funded by the Commission of European Communities, which aims to translate simultaneously from one of the commission's languages into the seven others, and 2) activities in Japan including projects to develop a computerised system for translating scientific and technical papers from English to Japanese and the reverse, a system for translating news items, a system designed to accept voice input and analyse meaning instead of just structure, and systems which use mini — and ultimately personal — computers rather than mainframes. Two basic factors are seen as responsible for the new push: developments in general data-processing technology and increasing demand for translation. The key barrier to the use of computers is seen as cost, but for large users there is evidence that machine translation does pay.

Keywords: **Eurotra project, Japan, machine translation, personal computer, speech understanding, voice recognition**

C17 Tanaka, H., Chiba, S., Kidode, M., Tamura, H. and Kodera, T.
Intelligent Man-Machine Interface
In (A11), pp.147–157

One of the goals of research into intelligent man/machine interfaces is to develop fundamental techniques which will afford flexible interactive facilities for Fifth Generation computer systems. The Japanese plan for research into intelligent man/machine interfaces is divided into three fundamental categories: 1) natural language processing; 2) speech processing; 3) picture and image processing. These researches will also help provide the foundation for the development of such basic applications systems as an intelligent question-answering system and a machine translation system.

Keywords: **image processing, intelligent interface, Japan, machine translation, natural language, speech understanding**

C18   Underwood, M. J.
      Intelligent User Interfaces
      In (A12), pp.135-143

This paper explores some of the difficulties and problems encountered in the establishment of intelligent user interfaces, with particular emphasis on the use of speech. Aspects of human communication and the difficulties of achieving machine understanding of human speech are assessed. The term 'speech understanding' was devised specifically to describe those machines that would infer the meaning of what was said in the manner of a human listener, without necessarily understanding every word. Machines of this kind are complex and some aspects of their design are considered. The Japanese have plans for making a significant improvement in speech understanding performance. The ambitious objectives of the Japanese project are compared with those of the Advanced Research Projects Agency (ARPA) funded programme in the United States which, five years after its completion, has not led to any speech understanding system in practical use.

Keywords: **FGCS project, intelligent interface, Japan, speech understanding, United States of America**

C19   Yasukawa, H.
      LFG in Prolog: Toward a Formal System for Representing Grammatical Relations
      ICOT Technical Report No. TR-019 (1983)

Currently ICOT plans the natural language understanding system focusing on discourse understanding. As the first stage of developing the system, the syntactic analysis modules are currently designed. Considering the requirements for the system, i.e. capability of treating the wide varieties of linguistical phenomena, a formal system for representing grammatical relations is needed in order to understand and maintain the grammar. Lexical Functional Grammar (LFG) proposed by Bresnan, Kaplan *et al.*, is considered as a candidate for the formal system for representing grammatical relations. The paper describes the experimental implementation of the LFG system in Prolog.

Keywords: **FGCS project, ICOT, Japan, LFG, natural language, Prolog**

D | Parallel Processing, Novel Architectures and VLSI

Abstract Numbers: D1-D56

See also: C2, C4, C5, C13, E12, E22, E24, G1

D1 Ackerman, W. B.
Data Flow Languages
Computer, Vol. 15, No. 2, pp.15-25 (February 1982).

The author explains the important properties of functional languages for parallel processing and compares them with imperative languages such as FORTRAN.

Keywords: **dataflow, functional programming, parallel processing**

D2 Aiso, H.
Fifth Generation Computer Architecture
In (A11), pp.121-127

This paper discusses the background and functional characteristics of Fifth Generation computers from the standpoint of computer architecture. The aim is to suggest reasonable approaches to architecture development and to offer major research subjects to be pursued for future computers, given the commitment to the development of very high-intelligence computer systems suitable for processing knowledge information. To this end it is intended to develop a logic programming language at the beginning of the project as an interface between hardware and software. The advanced architectures for the associated mechanisms such as inference machine, knowledge-base machine or intelligence interface machines will be investigated. Intelligent VLSI CAD systems and an integrated computer system to be used for designing VLSI chips and computer architectures will also be indispensable research subjects.

Keywords: **architecture, CAD, logic programming, VLSI**

D3 Allen, J.
Algorithms, Architecture, and Technology
In (A11), pp.277-281

With the advent of VLSI, the design of high-performance digital systems changes from the realisation of an algorithm on a given architecture fabricated in a fixed technology to an integrated process wherein the mutual interaction of algorithms, architecture and technology establishes a balanced design satisfying the system goals. This paper examines these factors in terms of the computational needs of Fifth Generation Computer Systems, with emphasis on applications in signal processing, speech generation and speech understanding.

Keywords: **architecture, signal processing, speech output, speech understanding, VLSI**

D4 Allen, J.
VLSI Overall System Design
In (A12), pp.31-39

This paper characterises the overall VLSI system design process in terms of a number of different tasks appropriate to a large variety of different representations. The functional tasks must be formally specified and the performance of each task must be varied over a large variety of architectural alternatives. The complexity of Fifth Generation designs will emphasise the need for a designer's expertise to be brought into programs that both generate these representations and transform between them. The three major VLSI technologies that are expected to gain prominence during this decade — CMOS, NMOS, and bipolar current mode logic such as ECL — are assessed.

Keywords: **VLSI**

D5 Amamiya, M., Hakozaki, K., Yokoi, T., Fusaoka, I., and Tanaka, Y
New Architecture for Knowledge Based Mechanisms
In (A11), pp.179-188

One of the most important research issues for Fifth Generation Computer Systems is the development of mechanisms for managing large amounts of knowledge data efficiently. This paper discusses the issue from the hardware implementation viewpoint. First, a definition of knowledge-base mechanism and its role in Fifth Generation Computer Systems is discussed in relation to the inference mechanism. Then, research relating to new architectures for knowledge-base mechanisms is described from the following viewpoints: 1) relational database machine architecture and its extension to knowledge-base machine, and 2) implementation of associative access control mechanisms in a knowledge-base memory system. Finally, the three-phase research and development plan is described.

Keywords: **architecture, inference, knowledge base, relational database**

D6 Arvind and Gostelow, K. P.
The U-Interpreter
Computer, Vol. 15, No. 2, pp.42-49 (February 1982)

This paper provides a description of how computations represented by cyclic dataflow graphs can be automatically unfolded to expose all parallelism to the underlying hardware.

Keywords: **dataflow, parallel processing, U-interpreter**

D7   Bell, D. H., Kerridge, J. M., Simpson, D. and Willis, N.
Parallel Programming: A Bibliography
Wiley Heyden Ltd (1983)

Parallel programming is where a program is constructed as two or more processes which work together to perform the required task. It is a technique which is likely to receive much wider attention in the computing community as a result of advances in microprocessor technology and programming languages. Over the years a large amount of research has been going on in the area and there is a large body of published literature on the subject. This bibliography provides a starting point for computer practitioners needing to familiarise themselves with the area. It provides a selective, rather than comprehensive, listing of around 350 major research papers, in easily accessible sources, concentrating particularly on programming. The papers are listed by major topic and a brief annotation of the contents of each paper is included. Work considered to be seminal or the first 'best paper' is starred to indicate where reading might begin in a particular subject area.

Keywords: **bibliography, parallel processing**

D8   Bell, D. H. and Simpson, D.
Think Parallel
Computer Bulletin, II/28, pp.20–21,25 (June 1981)

The authors put forward the argument that a natural view of the world is as a series of sequential processes which continue working in parallel until they need to co-operate to give and receive information and data. They show that such a view leads to a 'natural' way of writing programs which, further, makes the best use of available hardware resources. A program which performs text formatting is used as an example to illustrate the value of parallelism in contrast to the traditional sequential approach to programming.

Keywords: **parallel processing**

D9   Boral, H. and Dewitt, D. J.
Applying Data Flow Techniques to Data Base Machines
Computer, Vol. 15, No. 8, pp.57–63 (August 1982)
Although most research on dataflow languages and architectures sets out to produce a general purpose dataflow machine, the primary goal is generally to reduce the execution time of large numerical computations. In this paper, research into the application of dataflow machine principles towards improving access to large non-numerical databases is described.

The result of the authors' research into the problems of providing efficient access to relational databases that are too large to be handled by a single conventional processor is Direct — a multiprocessor, multiple instruction stream, multiple data stream, relational database machine. With an MIMD design such as Direct, groups of processors can work on different instructions from the same query, from different queries, or both. Some of the results obtained from experiments with various strategies for processor allocation for MIMD database machines are described, in particular those from variations of a strategy based on dataflow techniques.

Keywords: **architecture, dataflow, Direct, parallel processing, relational database**

D10 Cohen, C.
Fifth Generation Hardware Takes Shape
Electronics, Vol. 56, No. 15, pp.101-102 (July 28 1983)

This article reviews progress in the Japanese Fifth Generation Computer Systems project and the first year of operation of The Institute for New Generation Computer Technology (ICOT). It is reported that the hardware goals of the Fifth Generation Computer Systems project are in sight and that the pilot model of the special personal computer which will be used to develop software should be operational by the summer of 1984.

Keywords: **FGCS project, ICOT, Japan, personal computer**

D11 Davis, A. L. and Keller, R. M.
Data Flow Program Graphs
Computer, Vol 15, No. 2, pp.26-41 (February 1982)

This paper describes graphical dataflow languages and two different approaches to using graphs to represent dataflow programs — token models and structure models. It also explores the utility of graphs as a programming medium.

Keywords: **dataflow**

D12 Dennis, J. B.
Data Flow Supercomputers
Computer, Vol. 13, No. 11, pp.48-56 (November 1980)

The architects of supercomputers must meet three challenges if the next generation of machines is to find productive large-scale application to the important problems of computational physics. First, they must achieve high performance at acceptable cost; second, they must exploit the potential of LSI technology, and third, it must be possible to program supercomputers to exploit their performance potential. It is argued that present supercomputer architectures have exacerbated rather than resolved the software crisis.

Dataflow architectures are proposed as a way of solving the problem of efficiently exploiting concurrency of computation on a large scale while avoiding most of the difficulties which have hampered other approaches to highly parallel computation. Two dataflow-related architectures are presented: the dataflow multiprocessor and the 'cell block' architecture.

Keywords: **architecture, dataflow, parallel processing, supercomputers**

D13   Evans, D. J.
Parallel Processing Systems
Cambridge University Press (1982)

A collection of contributed papers based on an advanced course on parallel processing systems held at the University of Loughborough, England.

Keywords: **collection of papers, Loughborough University, parallel processing**

D14   Fairbairn, D. G.
VLSI: A New Frontier for Systems Designers
Computer, Vol. 15, No. 1, pp.87-96 (January 1982)

This paper presents a summary of those advances in VLSI technology that make highly parallel computation feasible, both for general purpose and for special purpose applications.

Keywords: **parallel processing, VLSI**

D15   Gajski, D. D., Padua, D. A., Kuck, D. L. and Kuhn, R. H.
A Second Opinion on Data Flow Machines and Languages
Computer, Vol. 15, No. 2, pp.58-69 (February 1982)

This paper presents a comprehensive critique of the entire dataflow approach and suggests that alternative approaches may offer more hope for the future.

Keywords: **dataflow**

D16   Galinski, C.
VLSI in Japan
Computer, Vol. 16, No. 3, pp. 14-21 (March 1983)

This paper offers a 'Pacific' viewpoint on the status of VLSI in Japan. It traces the historical origins of Japan's big leap forward in VLSI development, from the initial push towards a informatised society to the position in 1981 when Japan could boast 70% of the world market for 64K RAM chips using the latest VLSI technology. The author predicts, in conclusion, that the VLSI technology era will last until the end of the century impacting all production processes, especially integrated circuit fabrication and that by the end of the 1980s the Japanese computer industry will have closed the growth rate gap with United States industry. It is hoped that United States and European manufacturers can take advantage, in the future, of both the Japanese readiness for international co-operation and of the increase in the flow of information from Japan.

Keywords: **international relations, Japan, VLSI**

D17   Gurd, J. R.
Developments in Dataflow Architecture
In (A2)

Implicit in the Japanese Fifth Generation project proposals is the requirement for computer architectures to replace the traditional 'von Neumann' organisation. Dataflow architecture is one possible alternative which aims for high-speed computing via efficient exploitation of software parallelism in a highly parallel system of processing hardware. This paper describes separately the software and hardware aspects of dataflow architectures. In relation to software it considers the nature of software parallelism, the possible ways of representing it, implications for parallel machine-code design and techniques for compiling from various high-level languages into dataflow object-code. It considers the requirements for executing dataflow code and exploiting the exposed software parallelism. Three different system designs which have been or are being constructed as experimental research vehicles for further work applying and refining dataflow techniques are then described, and additional projects are briefly mentioned. Printed copies of the transparencies used by the author in his conference presentation are appended to the paper.

Keywords: **architecture, dataflow, parallel processing**

D18   Harkness, D. L. and Mills, A. F.
Data Flow Processing: A Review
Computer Bulletin, No. II/35, pp.14–17 (March 1983)

Data flow computers are considered to be an important building block for the Japanese Fifth Generation Computer project and data flow architectures have an advantage over other approaches to very high performance computing in that the scheduling and synchronisation of concurrent processes is handled at the hardware level. But to facilitate the expression of this concurrency requires a new approach to language design. The adoption of applicative (functional) languages may provide a solution, as could the use of data flow graphs as a primary programming language. This paper describes the operation of such graphs and looks at an alternative method of dataflow program representation, namely as a collection of activity templates, each of which corresponds to one or more nodes of a data flow graph. The architecture of a basic data flow computer is contrasted with the conventional von Neumann architecture and some of the problems that need to be solved in relation to data flow computers are raised.

Keywords: **dataflow, functional programming**

D19 Haynes, L. S., Lau, R. L., Siewiorek, D. P. and Mizell, D. W.
A Survey of Highly Parallel Computing
Computer, Vol. 15, No. 1, pp.9–24 (January 1982)

This survey describes classes of highly parallel machines (multiple special-purpose functional units, associative processors, array processors, dataflow processors, functional programming language processors and multiple CPUs), interconnection structures (cross-point and ring networks, systolic arrays, Banyan networks, the k-cube, cube-connected cycles and the perfect shuffle and tree structures), software development, and several application examples (artificial intelligence, Monte Carlo simulation, and partial differential equations).

Keywords: **artificial intelligence, dataflow, functional programming, parallel processing, partial differential equations, simulation**

D20 Horstmann, P. W.
Expert Systems and Logic Programming for CAD
VLSI Design, Vol. IV, No. 17 pp.37–50 (November 1983)

This article describes ways that expert systems based on the PROLOG language can solve real VLSI design problems. It is based on research at Syracuse University where logic programs for functional simulation, fault diagnosis and automatic test generation, and a rudimentary expert system in the area of design for testability have been developed.

Keywords: **automatic test generation, CAD, expert system, fault finding, Prolog, simulation, Syracuse University, VLSI**

D21 Hunt, D. J. and Reddaway, S. F.
Distributed Processing Power in Memory
In (A12) pp.49–62

High processing power can be achieved in a cost effective manner by having many processing units executing a common instruction stream and embedded in memory in order to provide a wide memory bandwidth. The ICL Distributed Array Processor uses this principle, the present implementation having 4096 elementary processors arranged in a square array. A parallel high-level language is used for expressing operations on arrays and permits effective use of the hardware's capability. Many applications have been implemented, including linear algebra, field equations, transforms, Monte Carlo methods, data mapping, sorting, table look-up, design automation, image processing, pattern recognition and symbol processing. Because of the bit organised nature of the processing elements, operations on low precision, symbol, and Boolean data give even higher performance than floating point work. The DAP is discussed in detail and applications are assessed.

Keywords: **Distributed Array Processor, distributed processing, ICL, image processing, mathematical application, parallel processing, pattern recognition, simulation**

D22 Ito, N., Onai, R., Masuda, K. and Shimizu, H.
Prolog Machine Based on the Data Flow Mechanism
ICOT Technical Memorandum No. TM-0007 (1983)

Prolog (Programming in Logic) is a simple but powerful language containing a basic inference capability. The execution process of Prolog is closely related to a dataflow concept. A model based on the data flow concept can naturally realise parallel processing. The authors are investigating a parallel Prolog machine based on the dataflow concept. This paper describes the basic idea of the parallel processing mechanism and the conceptual architecture of the machine.

Keywords: **architecture, dataflow, parallel processing, Prolog**

D23 Kakuta, T., Miyazaki, N., Shibayama, S., Yokota, H. and Murakami, K.
A Relational Database Machine 'Delta'
ICOT Technical Memorandum TM-0008 (1983)

This paper describes the following three items of a relational database machine named 'Delta': 1) Background and objective of its development, 2) Basic design architecture, and 3) The Prolog-based simulator for Delta commands.

Keywords: **Delta, Prolog, relational database**

D24 Kitsuregawa, M., Tanaka, H., and Moto-Oka, T.
Application of Hash to Data Base Machine and its Architecture
New Generation Computing, Vol. 1, No. 1, pp.63–74 (1983)

In this paper the application of the dynamic clustering feature of hash to a relational data base machine is discussed. By partitioning the relation using hash, large load reductions in join and set operations are realised. Several machine architectures based on hash are presented. A data base machine, GRACE, is proposed which adopts a novel relational algebraic processing algorithm based on hash and sort. Whereas conventional logic-per-track machines perform poorly in a joint dominant environment, GRACE can execute efficiently in time $O(N+M/K)$, where N and M are the cardinalities of two relations and K is the number of memory banks.

Keywords: **architecture, database machine, Grace, relational database**

D25 Kung, H. T.
Why Systolic Architectures?
Computer, Vol. 15, No. 1, pp.37–46 (January 1982)

This paper discusses the advantages of the systolic array approach to parallel computing. In systolic arrays, special purpose processors are connected in fixed topologies such as linear, mesh, and hexagonal nearest-neighbour arrays. It is argued that the approach can speed execution of computer-bound problems without increasing I/O requirements.

Keywords: **architecture, parallel processing, systolic array**

D26 Maller, V. A. J.
Content Addressing as an Aid to Information Management
In (A12), pp.89–100

Storage devices with content-addressable or associative properties have been discussed for many years, but although their utility has been acknowledged, the technology necessary to provide a device of anything more than trivial size has not been available. The main objective of the team at International Computers Ltd, Stevenage, who built the Content Addressable File Store, was to develop an 'attached processor' to which certain search and retrieval tasks could be physically devolved, but whose logical control would be administered by a mainframe resident data management system. Hardware, file organisation, performance and application features of the Content Addressable File Store are assessed.

Keywords: **Content Addressable File Store, ICL**

D27 Manuel, T.
Parallel Processing
Electronics, Vol. 56, No. 12, pp.105–114 (June 16 1983)

The serial nature of traditional computers is a major bottleneck for high-speed processing. Three types of design for faster processing are identified and research aimed at each of these is described. The first two involve, respectively, the use of: 1) an extremely fast single processor in a control-flow architecture and 2) a few very fast processors with the control flow enhanced with pipelines, vectoring, and some specialised processors. Many researchers believe, however, that the obvious answer is to go for the third option, parallel processing, but this necessitates radically new hardware and software architectures. The author of this special report on parallel processing argues that research on advanced parallel hardware and software architectures deserves and needs more support and attention to fulfil its promise. A number of parallel architecture projects are in high gear in five countries — United States of America, West Germany, France, Britain and Japan. A major part of this special report is an examination of some of these research projects, country by country, with attention focussed on hardware developments. The particular significance of data-flow architecture in parallel architecture research in all countries is highlighted.

Keywords: **architecture, control flow, dataflow, France, Japan, parallel processing, United Kingdom, United States of America, West Germany**

D28 Mead, C. A. and Conway, L. A.
Introduction to VLSI Systems
Addison-Wesley (1980)

A much referenced and often recommended textbook which introduces the techniques of designing VLSI chips. Topics covered include: devices, circuits, fabrication technology, logic design techniques and system architecture. The authors are careful to indicate what is dependent on today's technology and what is a principle of lasting importance.

Keywords: **VLSI**

D29 Mead, C. A. and Lewicki, G.
Silicon Compilers and Foundries Will Usher in User-Designed VLSI
Electronics, Vol. 55, No. 16, pp.107–111 (August 11 1982)

Reviews the rapidly changing state of the electronics industry and suggests appropriate routes for future development. The 'silicon foundry' processing chips to order is seen as inevitable. The high level of complexity of much modern semiconductor technology dictates a chip-design interface accessible directly to the user, with manual translation of circuit schematics being replaced by 'silicon compilers'. As a result, the next era in electronics is seen as being that of user-designed VLSI circuits and it is considered that the achievement of this should be given as much priority as the advancement of fabrication technology. Emphasis is given to the importance of building concurrent systems in the future since these will give an additional four orders of magnitude in computational capability compared with today's sequential systems. It is believed that there is no one highly concurrent architecture that serves all applications well; rather, classes of algorithms will be mapped onto suitable classes of concurrent structures, some broad but others very specific. Thus, architecture will not be separable from algorithms: programs and hardware will be one entity, 'entwined as a layout on the chip'.

Keywords: **architecture, VLSI**

D30 Murakami, K., Kakuta, T., Miyazaki, N., Shibayama, S., and Yokota, H.
A Relational Database Machine: First Step to Knowledge Base Machine
ICOT Technical Report No. TR-012 (1983)

The project to develop a knowledge base machine as part of the Japanese Fifth Generation Computer Systems is divided into three stages. In the first three-year stage, a working relational database machine is to be developed for the software development support system to be used in the second stage and also as an experimental system which provides a research tool for the knowledge base machine. The paper briefly describes the concepts and architecture of the relational database machine named 'Delta' which is currently under development at ICOT.

Keywords: **database machine, Delta, FGCS project, ICOT, Japan, knowledge base machine, relational database**

D31 Nishikawa, H., Yokota, M., Yamamoto, A., Taki, K. and Uchida, S.
The Personal Inference Machine (PSI): Its Design Philosophy and Machine Architecture
ICOT Technical Report No. TR-013 (1983)

As a software development tool of the Japanese Fifth Generation Computer Systems project, a personal sequential inference machine is now being developed. The machine is intended to be a workbench to produce a lot of software indispensable to the project. Its machine architecture is dedicated to effectively execute a logic programming language, named KL0, and is equipped with a large main memory, and devices for man-machine communication. It is estimated that its execution speed is about 20K to 30K LIPS. This paper presents the design objectives and the architectural features of the personal sequential inference machine.

Keywords: **architecture, FGCS project, Japan, KL0, logic programming, personal sequential inference machine**

D32 Randell, B. and Treleaven, P. C. (eds.)
VLSI Architecture
Prentice Hall International (1983)

This book provides a survey of VLSI design tools and techniques, and of current research on both general-purpose and special-purpose computer architecture suitable for VLSI implementation. It includes 28 papers given at the 1982 Advanced Course on VLSI Architecture held at the University of Bristol in July, sponsored by the UK Science and Engineering Research Council and the EEC CREST/ITG Committee. Contributions from industrial and academic researchers in the United Kingdom, Europe, America and Japan are included, with a series of six papers from the INMOS Corporation, Bristol, describing various aspects of their VLSI architecture design methodology.

Keywords: **architecture, Inmos, VLSI**

D33 Sakamura, K., Sekino, A., Kodaka, T., Uehara, T. and Aiso, H.
VLSI and System Architecture: The New Development of System 5G
In (A11), pp.189-208

This paper presents the R & D plan for VLSI CAD systems that are seen as indispensable for the construction of Fifth Generation computers and for a hardware environment called SYSTEM 5G on which the VLSI CAD systems run. SYSTEM 5G provides a network architecture involving super inference machines and many personal logic programming stations. The proposed CAD systems use a hierarchically organised design language in which anything from basic architectures of VLSI to VLSI mask patterns can be designed. Eventually the systems will be able to perform automatic design of VLSI chips when the characteristics of requirements of the chips is given. Details of the research and development plan and major research topics on SYSTEM 5G are discussed.

Keywords: **CAD, logic programming, networks, System 5G, VLSI**

D34 Shibayama, S., Kakuta, T., Miyazaki, N. and Yokota, H.
A Relational Database Machine 'Delta'
ICOT Technical Memorandum No. TM-0002 (1982)

Knowledge Base Machine is one of the expected achievements in the Fifth Generation Project. In the first-stage of the project, the primary object of KBM (Knowledge Base Machine) research is to provide a working Relational Database Machine (RDBM) named 'Delta', which serves multiple Sequential Inference Machine users via a local area network. Relational model was chosen as the working database model for the database machine because of its affinity to logic programming languages. Delta features: 1) High-Level Interface based on Relational Algebra, 2) Efficient Query Processing Capability by the use of dedicated Relational Database Engine (RDBE) Hardware, 3) Large Capacity Hierarchical Structured Memory, 4) Concurrent Query Processing enabling Multi-User Support, and 5) Recovery functions to keep consistency in the database. Command execution mechanism in the dedicated hardware (Relational Database Engine — RDBE) is based on sorting and merging of the attributes of relations transferred on-the-fly from the Hierarchical Memory Subsystem to RDBE, under control of the Control Processor. This paper describes the global architecture and concept of Delta. Brief descriptions of the subsystems which comprise Delta are also given.

Keywords: **database machine, Delta, FGCS project, Japan, knowledge base machine, logic programming, relational database**

D35 Shibayama, S., Kakuta, T., Miyazaki, N., Yokota, H. and Murakami, K.
On RDBM Delta's Relational Algebra Processing Algorithm
ICOT Technical Memorandum No. TM-0023 (1983)

Relational algebra processing and implementation overview of the database machine Delta, which is in the detailed design stage at ICOT, are described. In implementation design, existing components will constitute the database machine system. IP, CP consists of minicomputers, and RDBE consists of a piece of new hardware controlled by a minicomputer. HM consists of a general-purpose computer with large capacity moving-head disks. Data recovery functions in Delta needed in a practical environment are briefly described.

Keywords: **database machine, Delta, FGCS project, ICOT, Japan**

D36 Smith, K.
New Computer Breed Uses Transputers for Parallel Processing
Electronics, Vol. 56, No. 4, pp.67–68 (February 24 1983)

Computer researchers are beginning to break out of the serial processing straitjacket of von Neumann architecture by devising machines with a high degree of parallel processing achieved by combining discrete chunks of data and associated instructions into packets that can be processed simultaneously by a group of processor elements. This article describes one such computer, called Alice for Applicative Language Idealised Computing Engine, which is being developed at Imperial College, London. The prototype Alice will be a desktop unit capable of running fifth-generation languages as well as conventional software. It will be one of the first systems to use Inmos Corp's transputer, billed by the company as a fifth-generation building block. The article incorporates a note on the advantages of fifth generation languages.

Keywords: **ALICE, Imperial College, Inmos, parallel processing, transputer**

D37 Snyder, L.
Introduction to the Configurable, Highly Parallel Computer
Computer, Vol. 15, No. 1, pp.47–56 (January 1982)

This paper describes a machine which employs an array of general-purpose processors with interconnections that can be switched from one fixed topology to another by the action of an independent control function.

Keywords: **parallel processing**

D38 Stefik, M., Bobrow, D. G., Bell, A., Brown, H., Conway, L. and Tong, C.
Fifth Generation VLSI Design With AI
In (A2)

This paper proposes the use of explicit abstraction levels to organise decision making in digital integrated system design. These levels partition the concerns that a designer must consider at any time. Each level provides a vocabulary of terms and a set of simple composition rules which makes it possible to make a composition within a description level. To aid testing an expert system design, Palladio, is being developed which will assist a user to design with multiple levels of description. Palladio is being built in the knowledge engineering paradigm. It is argued that the use of multiple description levels provides more leverage for systematic exploration of possible designs than do silicon compilers or traditional register transfer hardware description languages. Printed copies of the transparencies used by the main author in his conference presentation are appended to the paper.

Keywords: **artificial intelligence, knowledge engineering, Palladio, VLSI**

D39 Tanaka, H., Amamiya, M., Tanaka, Y., Kadowaki, Y., Yamamoto, M., Shimada, T., Sohma, Y., Takizawa, M., Ito, N., Takeuchi, A., Kitsuregawa, M. and Goto, A.
The Preliminary Research on Data Flow Machine and Data Base Machine as the Basic Architecture of Fifth Generation Computer Systems
In (A11), pp.209-219

This paper summarises preliminary research on dataflow and database machines that will be used as the basis for the three-stage research subproject in this area that forms part of the Japanese Fifth Generation Computer Systems project. This preliminary work from July 1981 — February 1982 is concerned, first, with clarifying the research themes to be investigated and allocating them to the three stages and, second, selecting technological alternatives and suggesting some preliminary design specifications for the dataflow and database machines. Diagrams are included which give configurations for such machines.

Keywords: **database machine, dataflow, FGCS project, Japan**

D40 Treleaven, P. C.
Computer Architectures for the Fifth Generation
In (A12) pp.123-134

In their Fifth Generation Computer Project, the Japanese have taken the Artificial Intelligence view of the future generation of computers, while attempting to incorporate concepts from the research into processor designs to exploit VLSI, and novel parallel computers that efficiently support very high-level languages. This paper discusses the Fifth Generation Project in detail. Novel decentralised computers which might form the basis for a system architecture for Fifth Generation computers may be broadly classified as control flow, data flow, reduction, actor and logic computers. Finally, the possible future unification of four separate areas of current research — expert systems, very high-level programming languages, decentralised computing, VLSI technology — is discussed.

Keywords: **architecture, control flow, dataflow, distributed processing, expert system, FGCS project, Japan, VLSI**

D41 Treleaven, P. C.
Fifth Generation Computer Architecture Analysis
In (A11) pp.265-275 and in (A2)

The author argues that Fifth Generation Computer Systems will represent a unification of research into VLSI processors and into distributed processing. Each computer system will consist of a network of computing elements supporting an individual application or need. If all the special- and general-purpose VLSI processors are to be potential building blocks for larger systems, then it is necessary for them to conform to some overall program and machine organisation. Control flow, dataflow and reduction program organisations, and centralised, packet communication and expression manipulation machine organisations are examined with the aim of analysing their probable contribution to Fifth Generation computer architecture. Printed copies of the transparencies used by the author in his conference presentation are appended to the paper as it appears in (A2).

D42 Keywords: **architecture, control flow, dataflow, distributed processing, VLSI**

Treleaven, P. C. (ed.)
Proceedings of the Joint SRC/University of Newcastle upon Tyne Workshop on VLSI Machine Architecture and Very High Level Languages
University of Newcastle upon Tyne, Computing Laboratory Technical Report No. 156 (1980)

A collection of papers arising from a workshop held from 14-18 April 1980 to which leading researchers from the fields of the design of integrated circuits, parallel machine architecture and very high level programming languages were invited, with the aim of improving understanding of the goals and problems in the three fields.

Keywords: **architecture, CAD, collection of papers, functional programming, microcomputer, parallel processing, VLSI**

D43 Treleaven, P. C.
VLSI: Machine Architecture and Very High Level Languages
Computer Journal, Vol. 25, No. 1, pp.153-157 (February 1982)

This is a report on a VLSI workshop (organised by researchers at the University of Newcastle-upon-Tyne and sponsored by the UK Science Research Council) which was held in England on 14-18 April 1980. The workshop attracted 35 leading researchers from the US and Europe and the programme consisted of invited presentations interspersed with panel discussions. A brief summary of each of the sessions is given, together with details of how to obtain the complete proceedings. (*See* D42).

Keywords: **VLSI**

D44 Treleaven, P. C.
VLSI Processor Architectures
Computer, Vol. 15, No. 6, pp.33-45 (June 1982)

This paper presents a survey of novel VLSI processor architectures, which are based on a philosophy of simplicity and replication, and are leading the way towards a new generation of computers. The areas of VLSI processor architecture dealt with are special-purpose processors, simple microprocessors, tree machines, and non-von Neumann computers. It is concluded that if all VLSI processors discussed are to be used as building blocks in larger computer systems, they must conform to some common system architecture and many believe this necessitates a non-von Neumann architecture capable of exploiting both general- and special-purpose VLSI processors.

Keywords: **architecture, VLSI**

D45 Treleaven, P. C.
VLSI Processor Architectures
In (A2)

A new VLSI processor architecture 'culture' has rapidly developed, stimulated by Mead and Conway's 'Introduction to VLSI Systems' and by the companion multiproject chip courses and silicon foundries they have helped to establish. This paper is intended as a brief survey of these novel VLSI processor architectures which are implementable by only a few different types of simple cell, are characterised by simple and regular interconnections of cells, and use extensive pipelining and multiprocessing to achieve high performance. Examples range from identical chips configured to form special-purpose devices to replicated microcomputers that are interconnected to obtain parallel, non-von Neumann computers.

Keywords: **architecture, VLSI**

D46 Treleaven, P. C., Brownbridge, D. R. and Hopkins, R. P
Data-Driven and Demand-Driven Computer Architecture
Computing Surveys, Vol. 14, No. 1, pp.93–143 (March 1982)

Novel data-driven and demand-driven computer architectures are under development in a large number of laboratories in the US, Japan, and Europe. These are not based on the traditional von Neumann organisation; instead they are attempts to identify the next generation of computers. In data-driven (eg dataflow) computers, the availability of operands triggers the execution of the operation to be performed on them, whereas in demand-driven (eg reduction) computers the requirement for a result triggers the operation that will generate it.

This paper aims at identifying concepts and relationships that exist both within and between the two areas of research. Each architecture is examined at three levels: computation organisation, program organisation, and machine organisation. A survey of various novel architectures under development is also given.

Keywords: **architecture, data driven computer, dataflow, demand driven computer, Japan, United States of America, Western Europe**

D47 Treleaven, P. C., Hopkins, R. P. and Rautenbach, P. W.
Combining Data Flow and Control Flow Computing
Computer Journal, Vol. 25, No. 2, pp.207–217 (1982)

This paper presents a model of program organisation for a parallel, data-driven computer architecture which integrates the concepts of pure dataflow computation with those of 'multi-thread' control-flow computation. The model is presented at different levels of abstraction, including a discussion of the problems of program representation in data-driven computers and a logical description of a computer architecture which implements the organisation. The principal motivation behind the work described is to investigate how the concepts of dataflow and control flow interact.

Keywords: **control flow, dataflow**

D48 Turner, D. H.
Prospects for Non-Procedural and Data-Flow Languages
In (A2)

This paper discusses the likely impact on programming languages of the advent of large-scale parallelism in the hardware. It begins by outlining the properties of the 'von Neumann' tradition, both in languages and machines, which has been dominant to date, and the reasons why it is now being questioned. Then, the different styles of parallel architecture that have been proposed are briefly surveyed and the possible approaches to programming such architectures are considered. There are two main approaches: the first involves adding features for the explicit introduction and control of parallelism to relatively conventional sequential languages; the second, more radical approach (and it seems in the long run the only viable one) is to abandon programmer control of sequencing and to move to some kind of non-procedural language. Probably the most advanced type of non-procedural language currently available is the applicative language, based on lambda calculus or recursion equations. Examples from such a language, KRC (developed at the University of Kent), are given, and the future of such languages is considered. Printed copies of the transparencies used by the author in his conference presentation are appended to the paper.

Keywords: **dataflow, functional programming, KRC, parallel processing**

D49 Uchida, S.
Inference Machine: From Sequential to Parallel
ICOT Technical Report No. TR-011 (1983)

The inference machine planned for the Japanese Fifth Generation Computer Systems is scheduled to progress from a sequential inference machine to a parallel inference machine. The parallel inference machine is actually a parallel logic programming machine based on a parallel architecture such as dataflow machine. This paper outlines the language and architecture of a sequential inference machine, then discusses the various research themes that must be undertaken on the way to realizing a parallel inference machine and their current status. It also explains the future prospects for research aimed at developing parallel inference machines based on these present R & D themes.

Keywords: **architecture, dataflow, FGCS project, inference machine, Japan, logic programming, parallel processing**

D50 Uchida, S.
Toward a New Generation Computer Architecture
ICOT Technical Report No. TR-001 (1982)

This paper outlines the research and development plan for the Fifth Generation Computer Project from the viewpoint of computer architecture. The architectural goal of this project is to develop the basic technology to build a highly parallel processor which supports a logic programming language. As an intermediate goal, a parallel inference machine is considered. In the paper, an approach to the goal is introduced as well as motivations of the planning and its technological background.

Keywords: **architecture, FGCS project, inference machine, Japan, parallel processing**

D51 Uchida, S., Tanaka, H., Tokoro, M, Takei, K., Sugimoto, M. and Yasuhara, H.
New Architectures for Inference Mechanisms
In (A11), pp.167-178

This paper describes the three stages of the subproject of the 10-year research plan for the Japanese Fifth Generation Computer Systems concerned with the development of new architectures to implement inference mechanisms and the building of experimental machines powerful enough to fulfil the requirements of many applications in knowledge information processing. The inference mechanisms are given as a new programming language and its computational model, based on procedures of mechanical theorem proving in first order predicate calculus as seen in logic programming languages like PROLOG. The inference machine may thus be looked upon as a high-level language machine. To provide the machine with sufficient power, highly parallel architectures combined with VLSI technology are seen as indispensable. It is also expected that dataflow machines will become the basis of its architecture.

Keywords: **architecture, dataflow, FGCS project, Japan, inference, inference machine, knowledge engineering, logic programming, parallel processing, Prolog, VLSI**

D52 Uchida, S., Yokota, M., Yamamoto, A., Taki, K., and Nishikawa, H.
Outline of the Personal Sequential Inference Machine: PSI
New Generation Computing, Vol. 1, No. 1, pp.75-79 (1983)
Also available as ICOT Technical Memorandum No. TM-0005 (1983)

PSI (Personal Sequential Inference machine) is the first experimental model for the sequential inference machine supporting the logic programming language KL0 as part of the Japanese fifth generation computer project. This paper outlines the PSI architecture and hardware system.

Keywords: **FGCS project, Japan, KL0, logic programming, personal sequential inference machine**

D53 Watson, I. and Gurd, J.
A Practical Data Flow Computer
Computer, Vol. 15, No. 2, pp.51–57 (February 1982)

This paper describes the architecture and implementation of one experimental dynamic dataflow machine, which has eight unusual matching functions for handling incoming data tokens at its computational nodes.

Keywords: **dataflow**

D54 Werner, J.
Custom/Semicustom VLSI Design in Japan
VLSI Design, Vol. IV, No. 4, pp.30–50 (July/August 1983)

This article, based on the author's recent visit to Japan, describes the current custom and semicustom VLSI design work being done by leading Japanese companies.

Keywords: **Japan, VLSI**

D55 Werner, J.
The VLSI Connection in Two New Co-operative Research Programs
VLSI Design, Vol. IV, No. 1, pp.40–42 (January/February 1983)

This article describes two primarily United States programs which, although very different in focus and modus operandi, both deal with issues that vitally affect the country's ability to design and implement VLSI devices: Semiconductor Research Corporation (SRC) and Microelectronics and Computer Technology Corporation (MCC). The SRC will fund research efforts in universities in the form of individual grants or else by establishing 'centres of excellence', and currently has 13 participating companies. In contrast the MCC — once operational — plans to support research that the MCC itself will do, not via independent research contracts with universities. Task forces have been set up to zero in on four technology programs: microelectronics packaging, electronic computer-aided design and computer-aided manufacturing (CAD/CAM), advanced computer architecture, and software productivity. It is shown that the SRC is device-oriented, whereas the MCC is computer system oriented, but that there are some areas of overlap.

Keywords: **MCC, SRC, United States of America, VLSI**

D56 Yianilos, P. N.
A Dedicated Comparator Matches Symbol Strings Fast and Intelligently
Electronics, Vol. 56, No. 24, pp.113–117 (December 1 1983)

Until recently no single chip could perform intelligently comparisons between one string of characters and another. This article describes the PF474, from Proximity Technology Inc., which can carry out many such comparisons, and quickly. the PF474 uses statistical methods to identify a high degree of similarity between strings. Various applications for the chip are suggested, including the elimination of redundancy in databases such as mailing lists, searching very large databases and automatic speech recognition. The chip's notions of 'similarity' can be adjusted through software and the use of preprocessing of strings can increase its flexibility in use. The way in which the PF474's on-chip memory is mapped is described, together with its internal structure. The article concludes by considering how the chip's search power might be increased by exploiting parallelism. A note incorporated in the article offers a mathematical description of PF474's operation.

Keywords: **database, parallel processing, voice recognition**

E | Logic and Functional Programming

Abstract Numbers: E1-E37

See also: C8, C19, D1, D20, D22, D48, F24, F46

E1   Aida, H., Tanaka, H. and Moto-Oka, T.
A Prolog Extension for Handling Negative Knowledge
New Generation Computing, Vol. 1, No. 1, pp.87-91 (1983)

One of the defects of Prolog as a programming language is that it cannot easily handle negative information. This paper discusses a simple extension of Prolog based on the procedural interpretation of ordinary Prolog which makes it possible to deal with both positive and negative knowledge in a unified way and improves the terminability of programs. Utilisation of non-monotonicity to express knowledge such as default and inheritance, and the problems attached to the extension are also discussed.

Keywords: **Prolog**

E2   Backus, J.
Can Programming Be Liberated from the von Neumann Style? A Functional Style and its Algebra of Programs
Communications of the ACM, Vol. 21, No. 8, pp.613-640 (August 1978)

In this much-cited paper, the author argues that conventional programming languages have severe defects inherited from their common ancestor — the von Neumann computer. These include their division of programming into a world of expressions and a world of statements, their inability to effectively use powerful combining forms for building new programs from existing ones, and their lack of useful mathematical properties for reasoning about programs.

An alternative functional style of programming is described based on the use of 'combining forms' for creating programs.

Functional programs deal with structured data, are often non-repetitive and non-recursive, are hierarchically constructed, do not name their arguments, and do not require the complex machinery of procedure declarations to become generally applicable. Combining forms can use high-level languages to build still higher-level ones in a style not possible in conventional languages. Alternative computer designs to the von Neumann model are also considered, from the viewpoint of facilitating functional programming.

Keywords: **functional programming**

E3 Chikayama, T.
ESP — Extended Self-Contained PROLOG — as a Preliminary Kernel Language of Fifth Generation Computers
New Generation Computing, Vol. 1, No. 1, pp.11–24 (1983)
Also available as ICOT Technical Report No. TR-005 (1983)

In the first three-year development stage of the fifth generation computer systems project, a series of high-performance personal computers called sequential inference machines are being developed at ICOT Research Center. The machines have a high-level machine language called KL0, which is a PROLOG-based logic language with various extensions. In the software development of the sequential inference machines ESP, a software-supported yet higher level language compiled into KL0, is used instead of directly using KL0. This paper first describes the design principles of the total language system used in sequential inference machines including ESP and KL0. It also briefly explains KL0 design, as well as the design of various ESP features. The main extensions to PROLOG in KL0 are an extended control structure, multiple processes, operations with side-effects and hardware-oriented operations, while database management and name table management features are omitted.

Keywords: **ESP, FGCS project, ICOT, Japan, KL0, logic programming, personal sequential inference machine, Prolog**

E4 Clark, K. L.
An Introduction to Logic Programming
In (F45), pp.93–112

This paper introduces the use of the Horn clause subset of predicate logic as a programming language, mainly through example programs. The inference procedures and 'extra-logical' features of the logic programming language PROLOG are described and an indication given of how it might be used to implement Expert Systems.

Keywords: **expert system, logic programming, Prolog,**

E5 Clark, K. L. and McCabe, F. G.
PROLOG: A Language for Implementing Expert Systems
In (F32), pp.455–475
Also available as Imperial College of Science and Technology, Department of Computing, Technical Report DOC 80/21 (November 1980)

The authors briefly describe the logic programming language, PROLOG, concentrating on those aspects of the language that make it suitable for implementing Expert Systems. It is shown that features of Expert Systems such as inference-generated requests for data, probabilistic reasoning and explanation of behaviour are easily programmed in PROLOG. Each of these features is illustrated by showing how a fault-finder expert could be programmed in PROLOG. An appendix to the paper by P. Hammond describes a domain-independent Expert System that has been implemented in Micro-PROLOG and has been used with knowledge bases on car fault finding, skin disease diagnosis, ant identification, personal investment analysis and pipe corrosion diagnosis.

Keywords: **expert system, expert system shell, explanatory capability, fault finding, financial application, logic programming, medical application, Micro-Prolog, plausible reasoning, Prolog**

E6 Clark, K. L. and Tarnlund, S. A. (eds.)
Logic Programming
Academic Press (1983)

This book consists of twenty-three papers grouped into ten sections, each with its own theme, which include: introduction to logic programming, applications of logic programming, natural language understanding, implementation issues, specification and transformation, metalevel inference, control issues, logic programming languages, logic in LISP and Horn clause computability. The aim of these papers is to show how a view of computation as controlled inference can be exceedingly fruitful and leads naturally on to the idea that computers should be designed as inference machines: an idea that the Japanese have taken as the basis for their fifth generation machines.

Keywords: **collection of papers, inference machine, Lisp machine, logic programming, natural language**

E7 Clocksin, W. F. and Mellish, C. S.
Programming in Prolog
Springer-Verlag, 1981

This is the first generally available textbook with the aim of teaching PROLOG as a practical programming language. There are 11 chapters, entitled: tutorial introduction; a closer look; using data structures; backtracking and 'cut'; input and output; built-in predicates; more example programs; debugging PROLOG programs; using grammar rules; the relation of PROLOG to logic; projects in PROLOG.

The book contains many examples, all written in a 'core-PROLOG' which will run, with minor alterations, on any of the following four systems: DEC System 10 running TOPS-10, PDP-11 running UNIX, LSI-11 running RT-11 and ICL 2980 running EMAS. Appendices give details of language implementations on these and other systems.

Keywords: **logic programming, Prolog**

E8 Darlington, J., Henderson, P. and Turner, D. (eds.)
Functional Programming and its Applications
Cambridge University Press (1982)

This book is based on a course on functional programming and its applications held at the University of Newcastle-upon-Tyne in July 1980, and is aimed at professional programmers and researchers. It covers various functional languages and their use, methods of implementing functional languages, techniques of program development, mathematical foundations and special computer architectures for functional languages.

Keywords: **architecture, functional programming**

E9 Ferguson, R.
PROLOG: A Step Towards the Ultimate Computer Language
Byte, Vol. 6, No. 11, pp.384-399 (November 1981)

This article gives an introduction to the programming language PROLOG which the author considers is ideally suited to the manipulation of knowledge. Examples of PROLOG programs are given and a comparison with conventional procedural languages is made.

Keywords: **Prolog**

E10 Furukawa, K., Nakajima, R. and Yonezawa, A.
Modularization and Abstraction in Logic Programming
ICOT Technical Report No. TR-022 (1983)

In knowledge information processing, structuring of knowledge and algorithm is one of the key issues. The goal of this work is to introduce the concepts and mechanisms of abstraction, modularization and parameterization into logic programming which is one of the preliminary steps toward the kernel language of Fifth Generation Computer Systems.

Keywords: **FGCS project, knowledge engineering, logic programming**

E11 Goto, S.
Logic and Computers
ICOT Journal, No. 2 (September 1983)

This paper sketches some of the interrelations between logic and computations, and discusses PROLOG, the language emphasised by the Japanese Fifth Generation Computer Systems project, in terms of these relationships.

Keywords: **logic programming, Prolog**

E12 Gregory, S.
Parlog: A Parallel Logic Programming Language
Imperial College of Science and Technology, Department of Computing, Research Report (May 1983)

Parlog is a logic programming language in the sense that nearly every definition and query can be read as a sentence of predicate logic. It differs from Prolog in incorporating parallel modes of evaluation. For reasons of efficient implementation, it distinguishes and separates and-parallel and or-parallel evaluation. Parlog relations are divided into two types: and-relation and or-relation. A sequence of and-relation calls can be evaluated in parallel with shared variables acting as communication channels. Only one solution to each call is computed.

A sequence of or-relation calls is evaluated sequentially, but all the solutions are found by a parallel exploration of the different evaluation paths. A set constructor provides the main interface between and-relation and or-relation. This wraps up all the solutions to a sequence of or-relation calls in a list. The solution list can be concurrently consumed by an and-relation call.

The and-parallel definitions of relations that will only be used in a single functional mode can be given using conditional equations. This gives Parlog the syntactic convenience of functional expressions when non-determinism is not required. Functions can be invoked eagerly or lazily; the eager evaluation of nested function calls corresponds to and-parallel evaluation of conjoined relational calls.

This paper is a tutorial introduction and semi-formal definition of Parlog. It assumes familiarity with the general concepts of logic programming.

Keywords: **logic programming, parallel processing, Parlog**

E13 Hattori, T. and Yokoi, T.
Brief Constructs of the SIM Operating System
New Generation Computing, Vol. 1, No. 1, pp.81–85 (1983)
Also available as ICOT Technical Memorandum No. TM-0009 (1983), with the title "Basic Constructs of the SIM Operating System"

SIMPOS is an operating system for a super-personal computer (SIM), based on the logic programming language, KL0. This paper explains the basic constructs of SIMPOS — objects and relations, classes, pods and streams, worlds, process creation and operation, and ports and channels. It is anticipated that the entire system will be constructed using these basic constructs to ensure simplicity in concept and structure.

Keywords: **KL0, SIM, SIMPOS**

E14 Hikita, T.
Average Size of Turner's Translation to Combinator Programs
ICOT Technical Report No. TR-017 (1983)

D. A. Turner proposed in 1979 an interesting method of implementation for functional programs by first translating them to combinator expressions and then reducing the graphs they represent. J. R. Kennaway has recently shown that the worst case of the size of this translation is of order n^2 where n is the size of an original program. In this paper the average size of the translation is studied and it is shown that the order of the average is bounded by $n^{3/2}$. Results on lower bounds of the average are also shown in the case of programs with one variable. Finally, numerical statistics for the average size are exhibited, indicating that the size expansion is of reasonable range.

Keywords: **functional programming, graph reduction**

E15 Hirakawa, H.
Chart Parsing in Concurrent Prolog
ICOT Technical Report No. TR-008 (1983)

Current parsers such as DCG and BUP in logic programming languages are backtrack based parsers which are inefficient in that the same calculation along the different parsing paths is duplicated. To avoid this problem, the parser PCP is test developed, which is based on the parallel bookkeeping parsing method. PCP has the following features: 1) it handles grammar which includes left recursive and/or cycle rules; 2) the same subcalculation is only done once, and the result of the calculations is shared. PCP is implemented using the multi-process forks and the message passing mechanism of Concurrent Prolog.

Keywords: **Concurrent Prolog, logic programming, parallel processing, PCP**

E16 Hirakawa, H.
Implementing an OR-Parallel Optimizing Prolog System (POPS) in Concurrent Prolog
ICOT Technical Report No. TR-020 (1983)

This report discusses a computational model of an OR-Parallel Optimizing Prolog System (POPS) based on a graph-reduction mechanism and multi-processing, and its implementation in Concurrent Prolog. POPS executes Prolog in OR-Parallel and the same subcalculations are shared because of the graph-reduction mechanism. Furthermore POPS can handle the programs which include derivation cycles.

Keywords: **Concurrent Prolog, graph reduction, Prolog, POPS**

E17 Kowalski, R.
Logic as the Fifth Generation Computer Language
In (A12), pp.73–87

The Japanese Fifth Generation Computer Systems (FGCS) project has chosen logic as its core programming language. The target for 1990 can be summarised as the construction of a logic programming machine using highly parallel computer architecture based on VLSI technology for knowledge information processing applications. This paper describes the use of logic as a foundation for database query languages, together with the extra-logical features of PROLOG, and the use of logic as a computer language for children. The relationship between the use of logic for programming and its use for databases is illustrated clearly with the Horn Clause subset of logic. The procedural interpretation of Horn Clauses is identical to the problem reduction strategy in Artificial Intelligence, and problem reduction is discussed. The expressiveness of Horn Clauses can be improved by allowing implications as conditions. The use of PROLOG as a vehicle for the implementation of Expert Systems and functional and relational notation are examined.

Keywords: **database query system, expert system, logic programming, Prolog**

E18   Kowalski, R.
Logic for Expert Systems
In (F4), pp.79–93

In this paper, the author takes issue with those who identify expert systems with rule-based systems in general and those who identify them with the field of Artificial Intelligence as a whole. For this purpose, he presents a number of examples formulated as Horn clause programs and runnable in PROLOG. Some of the criticisms made of expert systems, in particular that they are subject to problems of correctness and maintainability, are also addressed and it is argued that the use of logic can help to solve these problems.

Keywords: **expert system, Prolog**

E19   Kowalski, R.
Logic for Problem Solving
North-Holland Publishing Co. (1979)

In this book the principles behind logic programming are introduced at a level suitable for the reader with no previous computing experience. The use of logic programming methods for solving a variety of problems is demonstrated and a comparison is made with conventional Artificial Intelligence problem-solving techniques.

Keywords: **logic programming**

E20   Kowalski, R.
Logic Programming in the Fifth Generation
In (A2)

The Japanese Fifth Generation Computer Systems project has chosen logic programming for its core programming language. The target for 1990 seems to be a logic programming machine exploiting highly parallel computer architecture using VLSI technology for knowledge information processing applications. This paper examines the general characteristics of PROLOG and logic programming, considers the reasons for adopting this approach and presents an introduction to the Horn clause subset of logic based on the author's experience with teaching the language to children. It is concluded that computer-executable logic is the missing link between functional programming languages being investigated for the exploitation of highly parallel computer architectures and the rule-based languages being used for applications in Expert Systems. It also provides the link needed to unify non-procedural specification languages and programming languages with query languages in database systems. Printed copies of the transparencies used by the author in his conference presentation are appended to the paper.

Keywords: **database query system, expert system, functional programming, logic programming, Prolog**

E21 Kunifuji, S. and Yokota, H.
PROLOG and Relational Databases for Fifth Generation Computer Systems
ICOT Technical Report No. TR-002 (1982)

This paper discusses the application of logic programming to relational databases. The authors describe how to interface between an extended PROLOG machine and an experimental relational algebra machine, and propose a simple way in which the PROLOG-like system can be modified to operate efficiently with large relational databases. To put it concretely, they propose an implemented example of such an interface module which is composed of three sub-modules, i.e. an access plan generator, an access plan converter, and a simulator of the relational database management system. In conclusion, they present a method to interface a logic programming language with a relational database management system.

Keywords: **logic programming, Prolog, relational database**

E22 Manuel, T.
Lisp and Prolog Machines are Proliferating
Electronics, Vol. 56, No. 22, pp.132–137 (November 3 1983)

This is the first part of a two part special report on the commercial status of artificial intelligence. It is concerned with the hardware systems now available and those under development around the world. It describes the personal Lisp machines currently available in the United States, as well as emerging general workstations and mainframe computers with AI language support. Research being carried out in Japan, the United Kingdom, and elsewhere is also described. Japanese developments include a machine which is not language restricted, a personal sequential-inference (PSI) computer for developing knowledge processing software and a parallel inference engine for a Prolog machine. In the United Kingdom, work is in progress on several forms of Prolog machine, ranging from a version of Prolog for Z80-based CP/M personal computers from Logic Programming Associates of London to a computer called Alice (Applicative language idealising computing engine) being developed at London's Imperial College. Alice will be a parallel-processing machine designed for

Parlog, a parallel version of Prolog, Lisp, and the college's own declarative fifth-generation language, Hope. The wide range of options available to those wishing to experiment with Prolog programming is described, together with research in the United States on advanced high-performance parallel AI architecture.

Keywords: **ALICE, architecture, artificial intelligence, Hope, Imperial College, inference, Japan, Lisp, Logic Programming Associates, parallel processing, Parlog, personal sequential inference machine, Prolog, United Kingdom, United States of America**

E23   Miyachi, T., Kunifufi, S., Kitakami, H., Furukawa, K., Takeuchi, S., and Yokota, H.
A Proposed Knowledge Assimilation Method for Logic Databases
ICOT Technical Memorandum No. TM-0004 (1983)

A relational database-oriented deductive query response system is considered as a logic database system. The authors propose a knowledge acquisition method suitable for such a system. As a knowledge acquisition method, the concept of knowledge assimilation oriented to deductive logic is formulated in an implementable form based on the notion of amalgamating object languages and meta languages. The concept of knowledge assimilation consists of checks on subconcepts called provability, contradiction, redundancy, independency, and corresponding internal database updates. Logic database-oriented knowledge assimilation was implemented in PROLOG, a logic programming language. As a result, PROLOG was found to be suitable for knowledge assimilation implementation, among other purposes.

Keywords: **knowledge acquisition, logic programming, Prolog, relational database**

E24   Myers, W.
Lisp Machines Displayed at AI Conference
Computer, Vol. 15, No. 11, pp.79–82 (November 1982)

This article describes some of the hardware and software displayed at the National Conference on Artificial Intelligence held in August 1982 at the University of Pittsburg and Carnegie Mellon University and sponsored by the American Association for Artificial Intelligence. The Lisp machines available from three major American companies — Lisp Machines, Inc., Symbolics, Inc., and Xerox — are described, together with some of the expert systems, natural language front ends, and other marketable systems that are emerging from Artificial Intelligence (AI). Consideration is also given to the reasons for the development of Lisp machines by the AI community and their value to those outside AI environments.

Keywords: **artificial intelligence, expert system, Lisp, natural language**

E25 Nakashima, H. and Suzuki, N.
Data Abstraction in Prolog/KR
New Generation Computing, Vol. 1, No. 1, pp.49-62 (1983)

As a means towards making Prolog suitable for writing large system software, the authors suggest the addition of features of object oriented languages. The paper describes the implementation of the features of data abstraction and inheritance in an expanded version of Prolog, Prolog/KR, how these can be used as abstract data types and how the hierarchy of these data types is introduced. The paper concludes by presenting an example of the use of an abstract data object in a simulation program of the memory system of a personal computer.

Keywords: **object oriented programming, Prolog, Prolog/KR, simulation**

E26 Robinson, J. A.
Logic Programming: Past, Present and Future
ICOT Technical Report No. TR-015 (1983)

This paper is an edited and condensed transcript of an ICOT Public Lecture given by the author in Tokyo on February 10, 1983. It describes the history of logic programming and gives the future of the idea of logic programming, distinguishing the following periods: Distant past (1879-1970), Near past (1971-1980) Near future (1991-2000) Distant future (2001-?).

Keywords: **logic programming**

E27 Robinson, J. A. and Sibert, E. E.
LOGLISP: An Alternative to PROLOG
In (F32), pp.399-419

For most people logic programming means PROLOG but it is not, and does not claim to be the definitive logic programming formalism and has some features which are alien to Kowalski's original conception of a deductive, assertional language based on the LUSH resolution of Horn clauses. The authors describe a system they have developed which seeks to create with LISP a faithful implementation of Kowalski's logic programming ideas.

Keywords: **logic programming, Loglisp, Prolog**

E28 Sato, M. and Sakura, T.
Qute: A Prolog/Lisp Type Language for Logic Programming
ICOT Technical Report No. TR-016 (1983)

Qute computes (partial) recursive functions on the domain S of symbolic expressions. Qute amalgamates Prolog and Lisp in a natural way. Any expression that is meaningful to Qute is either a Prolog expression or a

Lisp expression and a Prolog (or Lisp) expression is handled by the Prolog (or Lisp) part of Qute. Moreover, the Prolog and Lisp parts call each other recursively. Compared with the traditional Lisp symbolic expressions, the symbolic expressions in Qute are mathematically much neater and yet constitute a richer domain. Qute is a theoretically well-founded language defined on this domain of symbolic expressions. Qute has been implemented on VAX/UNIX and is used to develop a programming system for proving properties of symbolic expressions.

Keywords: **Lisp, logic programming, Prolog, Qute**

E29 Shapiro, E. Y.
A Subset of Concurrent Prolog and its Interpreter
ICOT Technical Report No. TR-003 (1983)

Due to its expressive power, simple semantics, and amenability to efficient implementation, Prolog is a promising language for a large class of applications. But Prolog has another, yet unexploited, aspect: it is a sequential simulation of a parallel computation model. There are two possible reasons for exploiting Prolog's underlying parallelism. One is to improve the performance of Prolog in some of its current applications, perhaps using novel computer architectures. The other is to incorporate in the range of Prolog applications those that require concurrency. Concurrent Prolog is concerned with both. Concurrent Prolog is a variant of Prolog currently under design by the author. This paper reports on a subset of Concurrent Prolog for which a working interpreter has been developed. The synchronization mechanism of Concurrent Prolog — read-only variables — can be viewed as a generalization of dataflow synchronization from functional to relational languages. The experience with implementing Concurrent Prolog, and the example programs in this paper suggest that this language may be a practical programming language for implementing operating system functions. Two controversial features of sequential Prolog, namely the cut and side-effects, are cleaned up in Concurrent Prolog. Concurrent Prolog's commit operator achieves an effect similar to cut in increasing the efficiency of the program, but it has a much cleaner semantics due to its symmetry, much the same way as Dijkstra's guarded-command has a cleaner semantics than the conventional if-then-else construct. Concurrent Prolog eliminates the need to use side-effects such as assert and retract to implement global data structures, since they can be implemented by perpetual processes, executing side-effect free programs.

Keywords: **architecture, Concurrent Prolog, dataflow, parallel processing, Prolog**

E30 Shapiro, E. and Takeuchi, A.
Object Oriented Programming in Concurrent Prolog
New Generation Computing, Vol. 1, No. 1, pp.25–48 (1983)
Also available as ICOT Technical Report No. TR-004 (1983)

Concurrent Prolog introduces an operational semantics of parallel execution to logic programs, thus allowing them to express concurrent computations. This paper focuses on the object oriented aspects of Concurrent Prolog and shows that the basic operations of creating an object, sending and receiving messages, modifying an object's state and forming class-superclass hierarchies can be implemented naturally in the language. A non-trivial Concurrent Prolog program — a multiple-window system — is studied in detail. The system is operational on the DECSYSTEM-20 and VAX-11 for a VT100 terminal. Traditional object-oriented programming is compared with object-oriented logic programming and two important programming techniques not easily available in the former are identified — incomplete messages and constraint propagation.

Keywords: **Concurrent Prolog, logic programming, object oriented programming, parallel processing**

E31 Sloman, A. and Hardy, S.
POPLOG: A Multipurpose Multi-language Program Development Environment
AISB Quarterly, No. 47, pp.26–34 (Spring/Summer 1983)

This article describes POPLOG, a multi-language system developed at the University of Sussex, combining POP-11 with LISP and PROLOG in an integrated, interactive program development environment. POPLOG provides a powerful yet 'friendly' environment for research, development and teaching in the field of Artificial Intelligence, but can also be used for more conventional programming. Its design is such that additional languages can be added and the system also includes a screen-editor, 'help' system and teaching system with a very large collection of documentation and library files. The system is now in use at several industrial and commercial organisations and a growing number of universities and colleges.

Keywords: **artificial intelligence, Lisp, Poplog, Prolog**

E32 Szeredi, P. and Sántáné-Tóth, E.
Prolog Applications in Hungary
In (A2)

This paper presents an overview of the main PROLOG applications in Hungary under the headings of pharmaceutical research, information retrieval, computer-aided design, software applications, supporting computer architecture design, simulation and other applications. For each application a short description of the problem and the main characteristics of the implementation are given. Finally, the paper analyses the reasons why PROLOG has been so successfully used for a relatively large number of applications in Hungary and some of the problems encountered. The transparencies used by the authors in their conference presentation are appended to the paper together with an extensive bibliography of articles and reports relating to PROLOG by Hungarian authors.

Keywords: **CAD, Hungary, information retrieval, medical application, Prolog, simulation**

E33   Takeuchi, A. and Furukawa, K.
Interprocess Communication in Concurrent Prolog
ICOT Technical Report No. TR-006 (1983)

Concurrent Prolog is a logic-based concurrent programming language which was designed and implemented on DEC-10 Prolog by E. Shapiro. In this paper, it is shown that the parallel computation in Concurrent Prolog is expressed in terms of message passings among distributed activities and that the language can describe parallel phenomena in the same way as Actor-formalism does. The expressive power of communication mechanism based on shared logical variables is examined and it is shown that the language can express both unbounded buffer and bounded buffer stream communication only by read-only annotation and shared logical variables. Finally a new feature of Concurrent Prolog is presented, which will be very useful in describing the dynamic formation and reformation of communication networks.

Keywords: **Concurrent Prolog, parallel processing**

E34   Tamaki, H. and Sato, S.
A Transformation System for Logic Programs Which Preserves Equivalence
ICOT Technical Report No. TR-018 (1983)

A program transformation system for logic programs based on the fold-unfold technique is defined and proved to preserve the equivalence of programs.

Keywords: **logic programming**

E35   Tamaki, H. and Sato, T.
Program Transformation Through Meta-Shifting
New Generation Computing, Vol. 1, No. 1, pp.93-98 (1983)

A program transformation strategy called meta-shifting is developed for control optimisation of programs written in languages with control freedom. The strategy involves writing a specialised evaluator for each program (segment) with specific control. It is applied to an equational language to derive a general method to transform one program into another, which simulates (approximately) normal order evaluation of the original program, under interpreters with arbitrary evaluation order.

Keywords: **program transformation**

E36 Warren, D. H. D.
Higher-Order Extensions to PROLOG: Are They Needed?
In (F32), pp.441–454

This paper examines two possible 'higher-order' extensions to the logic programming language PROLOG: 1) the introduction of lambda expressions and predicate variables so that functions and relations can be treated as 'first class' data objects; 2) the introduction of set expressions to denote the set of all (provable) solutions to some goal. Only the latter is seen as adding to the real power of the language, but, it is argued, must be defined with care.

Keywords: **logic programming, Prolog**

E37 Yokoi, T., Goto, S., Hayashi, H., Kunifuji, S., Kurokawa, T., Motoyoshi, F., Nakashima, H., Nitta, K., Sato, T., Shiraishi, T., Ueda, K., Umemura, M. and Umeyama, S.
Logic Programming and a Dedicated High-Performance Personal Computer
In (A11), pp.159–164

This paper describes research plans towards the development of a programming language and high performance personal computer as springboards for research and development in the area of Fifth Generation computer systems. The logic programming language PROLOG is to be taken as a starting point and this is to be extended so as to include valuable features of LISP. The developed language will first be implemented on a conventional large-scale computer and then on a high-performance personal computer. The specification of the machine to be developed is presented but it is stressed that the design must be flexible enough to enable future modification.

Keywords: **FGCS project, Japan, personal computer, Prolog**

F | Expert Systems and Artificial Intelligence

Abstract Numbers: F1-F67

See also: C8, D20, E5, E18

F1 **Artificial Intelligence: The Second Computer Age Begins**
Business Week, March 8 1982

This paper presents an overview of the characteristics of Expert Systems and the current scene in relation to commercially available products. The growing interest in America and other countries in the potential of research in AI is chronicled and evidence in the form of the establishment of industrial AI laboratories — most notably by Schlumberger Limited — and of commercial companies. The paper concludes by considering some of the social implications of intelligent computer systems, with particular reference to the potential threats to privacy and of 'AI unemployment'.

Keywords: **artificial intelligence, expert systems, Schlumberger Ltd, social implications, United States of America**

F2 **Artificial Intelligence Research in the Heuristic Programming Project**
AI Magazine, Vol. IV, No. 3, pp.81-92 (Fall 1983)

A brief account of the work currently in progress at the Stanford University Heuristic Programming Project (HPP), with pointers to more detailed published descriptions. The work is discussed under the following headings: the EURISKO project — a project that is investigating machine learning; the advanced architectures project; the knowledge acquisition project; the ONCOCIN project; the DART project; the NEOMYCIN/GUIDON2 project; the KBVLSI project; the MOLGEN project; the RX project; the IA project; the MRS project — a project that ultimately aims at constructing a program that can reason about and control its own problem solving activity; the blackboard architecture project; the GLISP project; the AGE project. In each case an indication is given of personnel involved in the project. A note on the HPP computing environment concludes the article.

Keywords: **Age, architecture, artificial intelligence, blackboard model, Dart, expert system, Eurisko, Glisp, Guidon, IA, knowledge acquisition, machine learning, Molgen, MRS, Neomycin, Oncocin, RX, Stanford Heuristic Programming Project, VLSI**

F3 **Digital and Artificial Intelligence**
Decnews, p.7 (October 1983)

A brief introduction to artificial intelligence and expert systems, and an account of the expert system tools currently available from Digital and their future plans in this area. An account of some of the co-operative activities involving Digital and the universities concludes the article.

Keywords: **artificial intelligence, Digital Equipment Corporation, expert system**

F4 **Expert Systems 83**
British Computer Society Specialist Group on Expert Systems (1983)

The proceedings of a conference held in Cambridge, England, in December 1983 by the British Computer Society Specialist Group on Expert Systems.

The proceedings comprise 25 papers subdivided into: experiences and case studies; continuous or real-time applications; expert system shells; expert system techniques; special interest sessions; expert system applications; knowledge acquisition; plus two invited talks (by Dr. A. Bundy and Professor R. Kowalski).

See also (E18) and (F18).

Keywords: **collection of papers, expert system, expert system shell, knowledge acquisition**

F5 **Heuristic Programming Project, 1980**
Stanford Heuristic Programming Project

This brochure describes the work of Stanford University Heuristic Programming Project which includes basic research in Artificial Intelligence, application-oriented research and the development of community tools useful for designing Expert Programs. Applications are described under the headings of science, medicine, engineering and education, together with a section on knowledge-engineering tools. A short introductory description of the concerns of the field of Artificial Intelligence is included in an appendix. Lists of project staff, doctoral dissertations produced and further reading are also included.

Keywords: **artificial intelligence, educational application, engineering application, expert system, expert system shell, knowledge engineering, medical application, scientific application, Stanford Heuristic Programming Project**

F6 Addis, T. R.
Expert Systems: An Evolution in Information Retrieval
Man-Computer Studies Group, Brunel University, Technical Report No. MCSG/TR14

This paper reviews the field of Expert Systems and concludes that the range of systems represents different degrees of enhancement to a basic computer information retrieval system. Extended relational analysis is introduced as a means of highlighting differences among systems. It is recommended that expert systems be viewed as information retrieval aids for experts and, thus, as a method of communication of group practices, rather than as simulated experts.

Keywords: **expert system, extended relational analysis, information retrieval**

F7 Addis, T. R. and Johnson, L.
Knowledge for Machines
In (A12) pp.17–30

The fifth generation computer is characterised by the change in view that 'knowledge' rather than 'data' is the essential raw material to be processed. It is commonly believed that by giving the computer knowledge, machine understanding could be improved. This paper examines both the abstract and technological aspects of knowledge. An approach to the abstract problem of defining knowledge is developed by distinguishing knowledge structures from meaning structures, and connecting knowledge with justification. The essentials of the three major knowledge representation schemes — semantic nets, production systems and frames — are examined. The conclusion is drawn that research is needed on the role of the users within a complete system. To construct a system with knowledge requires the ability to elicit knowledge and to construct an appropriate abstraction at a level of resolution adequate to deal with the set of expected situations for which the system was designed. New skills will have to be evolved and a new breed of computer expert ('knowledge engineer') will be required.

Keywords: **frame, knowledge acquisition, knowledge engineering, knowledge representation, production system, semantic net**

F8 Barnett, J. A.
Some Issues of Control in Expert Systems
Proceedings of the International Conference on Cybernetics and Society, Seattle, pp.1–5 (IEEE, 1982)

Because of the nature of the problem-solving domains in which expert systems operate, there are usually choices about what to do next. Many different rules are potentially useful to extend the current solution state. The control problem is to select the rule or rules that most likely will lead to acceptable behaviour. For many years, workers in artificial intelligence have tried to duck the control problem by compiling both control and application knowledge into the same large-grained rules. However, the desire to build systems that work in larger domains, fail safely at the boundaries of their competence, and are able to explain their behaviour, puts the importance of the control problem into sharp focus.

Keywords: **control problem, expert system**

F9   Barr, A. and Feigenbaum, E. A.
The Handbook of Artificial Ingelligence, Vol. 1
Pitman, London, 1981

First volume of a three-volume reference work on AI research conceived and produced at Stanford University's Department of Computer Science, with contributions from many other American universities and laboratories. It aims to introduce in a jargon-free way the important techniques and concepts of AI to computer technologists and users with no background in the area. It is arranged in five chapters, each comprising several items such as core papers on specific concepts, overviews, and accounts of particular AI programs. The first chapter presents a guide to the literature, and discusses the goals, history and current concerns of AI research. Chapters 2 and 3 introduce the key concepts associated with problem solving, search and knowledge representation. Chapters 4 and 5 describe research on 'natural languages' and the design of programs that understand spoken languages.

Keywords: **artificial intelligence, knowledge representation, natural language, problem solving, search, speech understanding**

F10   Basden, A.
On the Application of Expert Systems
International Journal of Man-Machine Studies, Vol. 19, No. 5, pp.461–477 (1983)

Expert systems have recently been arousing much interest in industry and elsewhere: it is envisaged that they will be able to solve problems in areas where computers have previously failed, or indeed, never been tried. However, although the literature in the field of expert systems contains much on their construction, on knowledge representation techniques, etc., relatively little has been devoted to discussing their application to real-life problems. This article seeks to bring together a number of issues relevant to the application of expert systems by discussing their advantages and limitations, their roles and benefits, and the influence that real-life applications might have on the design of expert systems software. Part of the expert systems strategy of one major chemical company is outlined. Because it was in constructing one particular expert system that many of these issues became important this system is described briefly at the start of the paper and used to illustrate much of the later discussion. It is of the plausible-inference type similar in operation to PROSPECTOR and its function is to predict the risk of stress-corrosion-cracking in stainless steels. The article is aimed as much at the interested end-user who has a possible application in mind as at those working in the field of expert systems.

Keywords: **expert system, plausible reasoning, scientific application**

F11   Bibel, W.
Logical Program Synthesis
In (A11) pp.227–236

Starting from an analysis of the reasons for the development of more flexible, more intelligent computer systems, this article elaborates some of their basic features. As programming is viewed as a paradigm for general problem solving, attention is focused on the various aspects of a future program synthesis system and the conceptual structure of a particular system called LOPS (Logical Program Synthesis system) is described. Because of its prominent role in LOPS, particular attention is devoted to the theorem proving method underlying the deductive component. The article concludes with some speculations on future developments and an account of technical requirements for future computer architecture.

Keywords: **architecture, LOPS, program synthesis, theorem proving**

F12 Bigger, C. J. and Coupland, J. W.
Expert Systems: A Bibliography
Institution of Electrical Engineers (1983)

This bibliography contains 121 references selected from the INSPEC database dealing with all aspects of expert systems, and covers the period 1969 onwards. It is divided into two sections: the first covers review articles and introductions to the subject, and the second deals with specific applications, subdivided into medical and other applications. Author and keyword subject indexes are included, together with a guide to journal abbreviations.

Keywords: **bibliography, expert system, medical application**

F13 Bird, J.
Expert Systems: The Facts Chase the Fiction
Micro Decision, No. 14, pp.99–102 (December 1982)

The author explains the state of current research into expert systems, and discusses one probable future role as human windows' to fifth generation computers.

Keywords: **expert system**

F14 Bramer, M. A.
A Survey and Critical Review of Expert Systems Research
In (F45), pp.3–29

This paper surveys the state of the art in Expert Systems, describing the major systems currently in existence and the underlying computational mechanisms. A number of associated theoretical issues are also discussed. References to papers on over 30 systems are included.

Keywords: **bibliography, expert system**

F15 Buchanan, B. G.
New Research on Expert Systems
In (F32), pp.269-299

The author begins by distinguishing Expert Systems from other AI programs in terms of their high performance, understandability and utility. He proceeds to survey the current state of the art and to consider directions for future work. The particular issues discussed that are likely to provide a focus for future work concern knowledge representation and control, explanatory capability, knowledge acquisition, validation, experimentation with existing AI systems and choosing a problem-solving framework.

Keywords: **expert system, explanatory capability, knowledge acquisition, knowledge representation, problem solving**

F16 Buchanan, B. G.
Partial Bibliography of Work on Expert Systems
Sigart Newsletter, No. 84, pp.45-50 (April 1983)

List compiled in November 1982 of over 200 references arranged alphabetically by author's surname. It includes items in journals, technical reports, and books.

Keywords: **bibliography, expert system**

F17 Bundy, A., Lumley, J., Merry, M and Sparck-Jones, K. (eds.)
A Catalogue of Artificial Intelligence Tools, Spring 1983
SERC Rutherford Appleton Laboratory (1983)

This catalogue seeks to promote interaction between members of the United Kingdom Artificial Intelligence (AI) community. It aims to do this by announcing the existence of AI techniques and portable software, and acting as a pointer into the literature. By AI techniques is meant algorithms, data (knowledge) formalisms, architectures, and methodological techniques which can be described in a precise way. The catalogue entries are intended to be non-technical and brief, but with a literature reference. By portable AI software is meant programming languages, shells, packages, toolkits, etc. which are available for use by AI researchers, including both commercial and noncommercial products. In every case details are given of where copies of the software may be obtained. There are two indexes: one, the 'logical table of contents', lists the entries under various subfields of AI; and one, the 'index of definitions', is a topic/keyword index. In addition there are some cross references amongst entries in the catalogue. The editors state that the catalogue has both a descriptive and a prescriptive role: it tries to say what AI is and what it should be. In particular it aims to 'cure' the methodological malaise in AI, which includes confusion over the nature of AI, over terminology and over how research might be evaluated. The current version of the catalogue is kept on line undergoing constant revision and refinement, and is to be printed periodically. (It is intended to produce a paperback version of the catalogue, aimed at the international AI community, which excludes entries only of interest in the United Kingdom context.)

Keywords: **artificial intelligence, catalogue of AI tools**

F18   Bundy, A., Silver, B. and Plummer, D.
An Analytical Comparison of Some Rule Learning Programs
In (F4), pp.184-223

This paper presents an analytical comparison of the rule learning programs of Brazdil, Langley, Mitchell et al, Shapiro and Waterman and the concept learning programs of Quinlan and Young. In order to clarify the similarities and differences amongst the techniques, they are described with uniform formalism and terminology. It is shown that the programs are tackling similar problems in similar ways. Some of the most useful and innovative techniques are pinpointed, together with a number of funnies in reported research.

Keywords: **knowledge acquisition, machine learning**

F19   Chandrasekaran, B. and Mittal, S.
Deep Versus Compiled Knowledge Approaches to Diagnostic Problem-Solving
International Journal of Man-Machine Studies, Vol. 19, No. 5, pp.425-436 (1983)

Most of the current generation expert systems use knowledge which does not represent a deep understanding of the domain, but is instead a collection of 'pattern → action' rules, which correspond to the problem-solving heuristics of the expert in the domain. There has thus been some debate in the field about the need for and role of "deep" knowledge in the design of expert systems. It is often argued that this underlying deep knowledge will enable an expert system to solve hard problems. In this paper the authors consider diagnostic expert systems and argue that given a body of underlying knowledge that is relevant to diagnostic reasoning in a medical domain, it is possible to create a diagnostic problem-solving structure which has all the aspects of the underlying knowledge needed for diagnostic reasoning 'compiled' into it. It is argued this compiled structure can solve all the diagnostic problems in its scope efficiently, without any need to access the underlying structures. Such a diagnostic structure is illustrated by reference to the authors' medical system MDX. The use of these knowledge structures in providing explanations of diagnostic reasoning is also analysed.

Keywords: **expert system, knowledge representation, MDX, medical application, problem solving**

F20 d'Agapeyeff, A.
Expert Systems, Fifth Generation, and UK Suppliers
NCC Publications (1983)

This report provides an introduction for the British computing supply industry to the areas of fifth generation computers and expert systems. It presents an outline description of the new software plus a short analysis of the relevant hardware developments for those previously unfamiliar with the areas of concern. Consideration is given to the nature of expert systems and their applications, the scope of the Japanese fifth generation computer systems project and the implications for British computer suppliers of the fifth generation, including some early market opportunities. Appendices present an annotated list of references for further reading and a list of relevant addresses.

Keywords: **expert system, FGCS project, Japan, United Kingdom**

F21 De Jong, K. A.
Knowledge Engineering: Where We Are and Where We Are Going
Proceedings of the International Conference on Cybernetics and Society, Seattle, pp.262–266 (IEEE, 1982)

The emergence of expert system technology has given rise to a new discipline called knowledge engineering. The Navy Centre for Applied Research in Artificial Intelligence has been evaluating the current state of this discipline in an attempt to assess its viability for transfer from the AI laboratory into selected application areas within the Navy. A summary of these observations and conclusions is presented.

Keywords: **expert system, knowledge engineering, Navy Centre for Applied Research in AI**

F22 Duda, R. O. and Gaschnig, J. G.
Knowledge-Based Expert Systems Come of Age
Byte, Vol. 6, No. 9, pp.238–281 (September 1981)

This paper considers how knowledge-based Expert Systems differ from other large computer programs written to solve special decision-making problems and explaining how Expert Systems work. In addition, several existing Expert Systems are briefly described and the operation of the mineral exploration program PROSPECTOR which the authors helped to develop is analysed. The paper concludes by presenting a small rule-based BASIC program for identifying animals which, although lacking many of the features of a 'real' Expert System, should serve as an effective teaching aid for those with a personal computer.

Keywords: **expert system, expert system teaching aid, personal computer, Prospector**

F23 Duda, R. O. and Shortliffe, E. H.
Expert Systems Research
Science, Vol. 220, No. 4594, pp.261–268 (April 15 1983)

Artificial Intelligence (AI), long a topic of basic computer science research, is now being applied to problems of scientific, technical and commercial interest. This article concerns that class of AI programs known as expert systems. The MYCIN system is described in detail, as a basis for making some general observations about expert systems. A number of other expert systems are briefly described to illustrate the themes identified. The article concludes by considering what problems need to be solved if the benefits of expert systems are to be realised in practice and by arguing that one of the major benefits of work in this area may be its impact on the systematisation and codification of knowledge previously thought unsuited for formal organisation.

Keywords: **artificial intelligence, expert system, knowledge representation, Mycin**

F24 Elcock, E. W.
How Complete are Knowledge-Representation Systems?
Computer, Vol. 16, No. 10, pp.114–118 (October 1983)

The author compares two related representation schemes for handling 'incomplete' knowledge, drawing analogies between Prolog (the most feasible of the first-order logic systems) and Absys (another assertive programming system developed at the University of Aberdeen in 1968). The comparisons are used to illustrate issues of incompleteness and the incompleteness resulting from any serious use of Prolog as a vehicle for a knowledge-based system is addressed.

Keywords: **Absys, incomplete knowledge, knowledge representation, Prolog**

F25 Feigenbaum, E. A.
Innovation in Symbol Manipulation in the Fifth Generation Computer Systems
In (A11), pp.223–226

Following a brief introduction, this text comprises the visual presentation materials used by the author in his talk. By this means, he presents the following arguments: 1) software rather than hardware problems are the critical ones for Fifth Generation systems; 2) knowledge, not inference, is the key to high levels of performance; 3) developments in software and knowledge acquisition necessitate major scientific innovations and innovations in the training and technical management of software engineers; 4) there must be a definite international commitment to innovation in the entire research and development process leading to the Fifth Generation Computer System.

Keywords: **international relations, knowledge based system**

F26 Feigenbaum, E. A.
Themes and Case Studies of Knowledge Engineering
In (F44), pp.3-25

This paper examines emerging themes of knowledge engineering, illustrates them with case studies drawn from the work of the Stanford Heuristic Programming Project and discusses general issues of knowledge engineering, art and practice. The major themes identified are generation-and-test, situation→action rules, domain-specific knowledge, flexibility to modify the knowledge base, comprehensibility of the line of reasoning, multiple sources of knowledge and explanatory capability. The major systems referred to are DENDRAL, META-DENDRAL, MYCIN-TEIRESIAS, SU/X, AM, PUFF, MOLGEN, and CRYSALIS.

Keywords: **AM, Crysalis, Dendral, explanatory capability, knowledge base, knowledge engineering, Meta-Dendral, Molgen, Mycin, Puff, Stanford Heuristic Programming Project, SU/X, Teiresias**

F27 Furukawa, K., Nakajima, R., Yonezawa, A., Goto, S. and Aoyama, A.
Problem Solving and Inference Mechanisms
In (A11), pp.131-138

The heart of the Fifth Generation computer is powerful mechanisms for problem solving and inference. The core of the computing system will be a deduction-oriented language. This paper describes the proposed R & D strategy for developments in this area. The preliminary version of this language will be based on the logic programming language, PROLOG, with three major extended features — modularisation mechanisms, metastructures and relational database inferences. Later versions will incorporate a parallel execution mechanism and specialised hardware architectures. The project will include an intelligent programming system, a knowledge representation language and system, and a meta-inference system built on to the core.

Keywords: **FGCS project, inference, problem solving, Prolog**

F28 Gevarter, W. B.
Expert Systems: Limited but Powerful
IEEE Spectrum, pp.39-45 (August 1983)

This article considers the range of tasks that may be carried out by an expert system, describes how such systems are structured, and how knowledge is represented in them, gives an account of some well-known systems — Dendral, Mycin, Internist, and R1 — and of a hypothetical system for space shuttle flight operations, examines the range of architectures of expert systems, and considers the limitations of such systems. It is concluded that, despite the many limitations that can be identified, there have been some notable successes and there appear to be few constraints on the ultimate use of expert systems. The article includes a table of existing expert systems and concludes with a list of further reading.

Keywords: **architecture, Dendral, expert system, Internist, Mycin, R1**

F29 Gevarter, W. B.
An Overview of Expert Systems
Proceedings of the International Conference on Cybernetics and Society, Seattle, pp.156–160 (IEEE, 1982)

This article provides an overview of expert systems. Topics covered include: what an expert system is, techniques used, existing systems, applications, major participants, the state of the art, research requirements, and future trends and opportunities.

Keywords: **expert system**

F30 Harris, L. R.
Fifth Generation Foundations
Datamation, Vol. 29, No. 7, pp.148–156 (July 1983)

The purpose of this article is to describe the Japanese fifth generation project in terms of some current approaches to data processing, in order to impart some understanding of what the Japanese are trying to achieve and what their chances of success really are. The five topic areas of knowledge bases, knowledge base query, the inference machine, natural language query and expert systems are explained and their relationship to current systems is examined. Each step in the project is shown to be based on existing capabilities. It is concluded that, while it is almost certain that the Japanese will not attain their project goals, they will fulfil some of them. The issue of problem selection is seen as a particularly significant area. It is argued that many of the problems selected by the Japanese are unlikely to be solvable using the expert system approach in the time frame of the project but an interesting question will be what class of problems will be brought into the window of expert system technology by improved hardware.

Keywords: **expert system, FGCS project, inference machine, Japan, knowledge base, natural language**

F31 Hawkins, D.
An Analysis of Expert Thinking
International Journal of Man-Machine Studies, Vol. 18, No. 1, pp.1–47 (January 1983)
Also available as Schlumberger-Doll Research Technical Report (July 1983)

In the belief that human expertise should be better understood before the users of Expert Systems specify the services needed and expected from such systems, this paper presents an analysis of expert thinking particularly with reference to the area of petroleum geology. It begins by giving examples of the services users obtain from human experts in the area, as a means of indicating the general requirements desirable in a human and, by analogy, an Expert System. A theory of expert thinking is then developed which identifies a number of human knowledge-handling techniques which could be implemented in a system. Finally, some different types of user interaction are examined to distinguish areas where a system might succeed or fail in understanding its users' needs.

Keywords: **expert system, geological application, human knowledge processing**

F32   Hayes, J. E., Michie, D. and Pao, Y-H. (eds.)
Machine Intelligence 10
Ellis Horwood (1982)

The most recent in a celebrated series of collected papers on machine intelligence, this volume contains 29 papers on the topic of 'Intelligent Systems: Practice and Perspective'. These are divided into nine sections, namely: mechanised reasoning, reasoning about computations, acquisition and matching of patterns, computer vision, problems of robotics, knowledge-based systems, logic programming, programs of the brain, philosophies of man and machine.

(See also E5, E27, E36; F15, F50, F54)

Keywords: **artificial intelligence, collection of papers, knowledge based system, logic programming, pattern acquisition, pattern matching, reasoning, robotics, vision**

F33   Hayes-Roth, F., Waterman, D. A. and Lenat, D. B. (eds.)
Building Expert Systems
Addison-Wesley (1983)

This book arose out of a workshop on expert systems which brought together expert system researchers and developers with the aim of synthesising the knowledge of the field. It is designed to provide a broad introduction to the concepts and methods necessary for an understanding of how expert systems work. Chapters are grouped into five sections: introduction, building an expert system, evaluating an expert system, expert system tools, and a typical problem for expert systems. Exercises appear at the end of each chapter and an extensive bibliography is included. It is intended that the book should serve both as a student text and as a survey for scientists, engineers, and technical managers.

Keywords: **bibliography, expert system**

F34   Jordan, J. A. Jr., Hirsch, P. M., Hollander, C. R. and Reinstein, H. C.
Expert Systems and Technology Transfer
British Computer Society Specialist Group on Expert Systems Newsletter, No. 7, pp.16-23 (January 1983)

Reprinted from the Proceedings of the 1982 South East Asia Regional Computer Conference

Every society is challenged to adapt and integrate appropriate new technologies. The computer is both a new technology and a tool by which knowledge of new technologies may be transferred to non-experts. Expert systems deliver knowledge to help non-experts solve specific problems. The features of expert systems are illustrated through an account of DART, an experimental fault diagnosis system which uses the EMYCIN expert systems software, and EXES/370, a system producing system flow charts from minimal dialogue with application experts. The challenges which need to be overcome before expert system technology can enjoy wide applicability are discussed, together with the potential for technology transfer of expert systems. It is concluded that, if the potential of expert systems can be exploited, developing countries will be able to develop more rapidly with fewer mistakes while developed ones will be able to use their technologies to better advantage.

Keywords: **Dart, EXES/370, expert system, technology transfer**

F35 Kahn, G.
The Scope of Symbolic Computation
In (A11), pp.237–242

This paper examines the scope of symbolic computation. To do this it first reviews some of the more interesting systems in existence today, including the MACSYMA system for calculus, syntax systems, the program manipulation system MENTOR, the semantic definition system SIS, the proof-checking system LCF and the theory compiler FORMEL. It goes on to explain some methods and problems in this area, in particular the efficiency problems that often arise. Finally some future research areas are identified and their implications for computer architecture considered. The areas mentioned include VLSI design, document preparation systems, and computer-aided instruction.

Keywords: **CAI, document preparation, Formel, LCF, Macsyma, Mentor, SIS, symbolic computation, VLSI**

F36 Kitakami, H., Funifuji, S., Miyachi, T., and Furukawa, K.
A Methodology for Implementation of a Knowledge Acquisition System
ICOT Technical Memorandum TM-0024 (1983)

This paper describes an investigation conducted on the Knowledge Acquisition System for a Knowledge Base System, and discusses a conceptual configuration and implementation method for some mechanisms in this system. These mechanisms include a meta inference mechanism, inductive inference mechanism, knowledge assimilation mechanism and knowledge accommodation mechanism. These mechanisms enable this system to turn the knowledge into the user's purpose. The discussion of the implementation method for the inductive inference mechanism attempts to explain speeding up strategy: these mechanisms include the manipulation of facts, rules and integrity constraints as knowledge.

Keywords: **inference, knowledge acquisition, knowledge based system**

F37 Langlotz, C. P. and Shortliffe, E. H.
Adapting a Consultation System to Critique User Plans
International Journal of Man-Machine Studies Vol. 19, No. 5, pp.479–496 (1983)

A predominant model for expert consultation systems is one in which a computer program simulates the decision making processes of an expert. The expert system typically collects data from the user and renders a solution. Experience with regular physician use of ONCOCIN, an expert system that assists with the treatment of cancer patients, has revealed that system users can be annoyed by this approach. In an attempt to overcome this barrier to system acceptance, ONCOCIN has been adapted to accept, analyse, and critique a physician's own therapy plan. A critique is an explanation of the significant differences between the plan that would have been proposed by the expert system and the plan proposed by the user. The critique helps resolve these differences and provides a less intrusive method of computer-assisted consultation because the user need not be interrupted in the majority of cases — those in which no significant differences occur. Extension of previous rule-based explanation techniques has been required to generate critiques of this type.

Keywords: **consultation system, expert system, explanatory capability, Oncocin**

F38 Lowrance, J. D. and Garvey, T. D.
Evidential Reasoning: A Developing Concept
Proceedings of the International Conference on Cybernetics and Society, Seattle, pp.6–9 (IEEE, 1982)

One common feature of most knowledge-based expert systems is that they must reason based on evidential information. Yet there is very little agreement on how this should be done. The authors present their current understanding of this problem and some partial solutions. They begin by characterising evidence as a body of information that is uncertain, incomplete and sometimes inaccurate. Based on this characterisation, the authors conclude that evidential reasoning requires both a method for pooling multiple bodies of evidence to arrive at consensus opinion and some means of drawing the appropriate conclusions from that opinion. This approach, based on a relatively new mathematical theory of evidence, is contrasted with those approaches based on Bayesian probability models. The authors believe that their approach has some significant advantages, particularly its ability to represent and reason from bounded ignorance.

Keywords: **evidential reasoning, expert system**

F39 McCalla, G. and Cercone, N.
Approaches to Knowledge Representation
Computer, Vol. 16, No. 10, pp.12-18 (October 1983)

This paper provides both an introduction to the fifteen articles brought together in this special issue of the journal on knowledge representation, and some background and context to the articles by mapping out the basic approaches to knowledge representation that have developed over the years. The most important current approaches are described, namely semantic networks, first-order logic, frames and production systems. However, it is noted that some approaches to representation, such as analogical representations, do not fit neatly into these categories and that many issues are common to all approaches. Some current research themes in the area are discussed, in particular how to define the precise dimensions and formal underpinnings of knowledge representation, handling imprecise information, knowledge acquisition and building practical systems. Brief summaries of all the articles in the special issue are included in the introduction.

(See also (F24) and (F66))

Keywords: **collection of papers, frame, knowledge acquisition, knowledge representation, plausible reasoning, production system, semantic net**

F40 Manuel, T. and Evanczuk, S.
Commercial Products Begin to Emerge from Decades of Research
Electronics, Vol. 56, No. 22, pp.127-129 (November 3 1983)

The first wave of computer products based on Artificial Intelligence (AI) technology are beginning to appear after a long history of international research. This article takes the view that the expert and natural language systems that have recently come into use could herald a tidal wave of such products in the future. Topics discussed include the range of definitions of AI, the developments in hardware languages and program-development environments that make AI work more feasible and initiatives in Japan, the United States and Europe to promote AI research and development.

Keywords: **artificial intelligence, expert system, Japan, natural language, United States of America, Western Europe**

F41 Michaelsen, R. and Michie, D.
Expert Systems in Business
Datamation, Vol. 29, No. 11, pp.240-246 (November 1983)

This article begins by contrasting expert systems with the long available decision support systems, arguing that the former are concerned with decisions that are not fully understood. Hence, expert systems require different programming techniques, most notably methods based on production systems, to ensure easy program revision and the possibility of the system explaining its line of reasoning. A list of successful expert systems in the public domain is presented and it is suggested that the lack of business applications is probably accounted for by behavioural variables that can slow down acceptance of such systems. Nevertheless, it is argued that expert systems should prove successful in business environments because they involve a type of decision making — using rules of thumb to make decisions in complex domains — that is common in business. A number of business systems under development are identified, which offer assistance to auditors and tax experts. TAXADVISOR, a system developed by the authors to interact with a human tax consultant to give estate planning tax advice for clients, is described in some detail as a means of dispelling some of the myths about expert systems. The article concludes by identifying a number of business areas where expert systems might be developed and by offering an optimistic view of the future prospects of expert systems in business.

Keywords: **business application, decision support system, expert system, production system, Taxadvisor**

F42 Michalski, S., Carbonell, J. G. and Mitchell, T. M. (eds.)
Machine Learning: An Artificial Intelligence Approach
Tioga Publishing Co., Palo Alto, California (1983)

This book contains tutorial overviews and research papers representative of contemporary trends in the area of machine learning as viewed from an artificial intelligence perspective. The individually authored chapters are entitled: An overview of machine learning; Why should machines learn?; A comparative review of selected methods for learning from examples; A theory and methodology of inductive learning; Learning by analogy: formulating and generalizing plans from past experience; Learning by experimentation: acquiring and refining problem-solving heuristics; Acquisition of proof skills in geometry; Using proofs and refutations to learn from experience; The role of heuristics in learning by discovery: three case studies; Rediscovering chemistry with the BACON system; Learning from observation: conceptual clustering; Machine transformation of advice into a heuristic search procedure; Learning by being told: acquiring knowledge for information management; The instructable production system: a retrospective analysis; Learning efficient classification procedures and their application to chess endgames; Inferring student models for intelligent computer-aided instruction. In addition, the volume includes a comprehensive bibliography of machine learning and a glossary of selected terms in machine learning.

Keywords: **artificial intelligence, automatic induction, Bacon, bibliography, chess, collection of papers, machine learning, production system, scientific application**

F43 Michie, D.
Aspects of the Fifth Generation: The Japanese Knowledge Bomb
In (A2)

This paper reviews the Japanese Fifth Generation proposals, principally from an Artificial Intelligence viewpoint. It is argued that Artificial Intelligence has a crucial bridging function in the Japanese proposals, namely to mediate between a representation of a user's problem and its solution in an extremely high-level language running on a Fifth Generation machine, to produce a solution in a form acceptable and intelligible to the user. Particular attention is focused on recent work on automatic induction, particularly on so-called 'structured induction'. The need for a scientifically sound theory of measuring the knowledge content in a knowledge based system is argued.

There are two appendices: the first contains short notes on those of the 26 themes listed in (A5) which the author believes are the most appropriate for special effort in the UK; the second is a prospectus for 'The Turing Institute' — a proposal for a national research institute for computer science and engineering.

Printed copies of the transparencies used by the author in his conference presentation are appended to the paper.

Keywords: **artificial intelligence, automatic induction, FGCS project, Japan, Turing Institute, United Kingdom**

F44   Michie, D.(ed.)
Expert Systems in the Micro-Electronic Age
Edinburgh University Press (1979)

This book is a collection of the 18 papers of the 1979 AISB Summer School on Expert Systems. The papers deal with the principles behind such systems, their logical basis, their control components and rule-based structure, as well as with such specific programs as MECHO (for solving problems in mechanics), CONGEN (aiding the structural chemist), PROSPECTOR (a consultant system for mineral exploration) and RITA (an intelligent interface package).

(See also F26)

Keywords: **collection of papers, Congen, expert system, Mecho, Prospector, Rita**

F45   Michie, D. (ed.)
Introductory Readings in Expert Systems
Gordon and Breach (1982)

This is a collection of 12 papers on Expert Systems with the following titles: 'A survey and critical review of Expert Systems research'; 'Fundamentals of the knowledge engineering problem'; 'PROSPECTOR: an Expert System for mineral exploration'; 'Research in office semantics'; 'Fundamentals of machine-oriented deductive logic'; 'An introduction to logic programming'; 'Chess endgame advice: a case study in computer utilisation of knowledge'; 'Problems and trends for the future of logic programming'; 'Intelligent systems in education'; 'Streamlining problem-solving processes'; 'Semi-autonomous acquisition of pattern-based knowledge'; 'The state of the art in machine learning'.

(See also E4; F14, F48, F50)

Keywords: **chess, collection of papers, educational application, expert system, knowledge acquisition, knowledge engineering, logic programming, machine learning, problem solving, Prospector**

F46 Mizoguchi, F.
PROLOG Based Expert System
New Generation Computing, Vol. 1, No. 1, pp.99–104 (1983)

A PROLOG based expert system called APLICOT is described. Its basic design framework is the same as used for other languages implementing expert systems. This utilisation of the same design framework permits a comparison to be made of system performance obtained within the same domain of knowledge. The domain chosen involves a reactor's fault diagnosis system consisting of 76 production rules. The results of this comparison show the overall performance of APLICOT to be of the same level as that of EXPERT, EMYCIN, and ADIPS, which were developed in the past utilising different programming languages. Although its code size is ten times smaller than that of LISP or FORTRAN based expert systems, APLICOT's backward and forward reasoning system gives it the same level of system performance and flexible inference strategy as these other systems, suggesting the potential software productivity of PROLOG based expert systems.

Keywords: **Adips, Aplicot, Emycin, Expert, expert system, expert system shell, fault finding, Prolog**

F47 Nau, D. S.
Expert Computer Systems
Computer, Vol. 16, No. 2, pp. 63–85 (February 1983)

Most expert computer systems organise knowledge on three levels: data (representing declarative knowledge), knowledge base (domain-specific problem-solving knowledge), and control (procedural knowledge control strategies). This article discusses the techniques used in expert systems on each of these levels. In addition information is included on AI problem-solving and knowledge-representation techniques. Examples are drawn from a wide range of systems.

Keywords: **artificial intelligence, expert system, knowledge base, knowledge representation, problem solving**

F48 Quinlan, J. R.
Fundamentals of the Knowledge Engineering Problem
In (F45), pp.192–207

This paper deals with two of the central problems in the design of Expert

Systems: 1) how is the knowledge to be represented, so that it is both usable by the system and comprehensible by human beings?; 2) what architecture should the system have so that the knowledge can be brought to bear on problems in this domain?

The general ideas are illustrated with a sampler of current systems. Three methods of encoding knowledge — production systems, first-order logic statements and frame systems — are discussed, together with two current architectures for deploying knowledge — inference engine and the blackboard model. Brief consideration is also given to the problem of how knowledge is acquired, with particular reference to ways of automating the process.

Keywords: **architecture, blackboard model, expert system, frame, inference engine, knowledge acquisition, knowledge representation, production system**

F49 Quinlan, J. R.
Inferno: A Cautious Approach to Uncertain Inference
Computer Journal, Vol. 26, No. 3, pp.255-269 (August 1983)

Expert systems commonly employ some means of drawing inferences from domain and problem knowledge, where both the knowledge and its implications are less than certain. Methods used include subjective Bayesian reasoning, measures of belief and disbelief, and the Dempster-Shafer theory of evidence. Systems based on these methods are analysed in this paper and important deficiencies are identified, in areas such as the reliability of deductions and the ability to detect inconsistencies in the knowledge from which deductions were made. A new system called INFERNO is described which addresses some of these problems. Its two major contributions are the guaranteed validity of any inference that it makes and its concern for, and assistance in establishing the consistency of information about the problem and its domain. The use of INFERNO is illustrated and compared with one of the more powerful Bayesian systems in common use, AL/X. The paper concludes by evaluating INFERNO's contribution to inference under uncertainty and suggesting directions for further work.

Keywords: **AL/X, Bayesian system, expert system, inference, Inferno, plausible reasoning**

F50 Quinlan, J. R.
Semi-Autonomous Acquisition of Pattern-Based Knowledge
In (F32), pp.159-172 and in (F45), pp.192-207

This paper has three themes: 1) the task of acquiring and organising knowledge on which to base an Expert System is difficult; 2) inductive inference systems can be used to extract this knowledge from data; 3) the knowledge so obtained is powerful enough to enable systems using it to compete handily with more conventional algorithm-based systems.

These themes are explored in the context of attempts to construct high-performance programs relevant to the chess endgame king-rook versus king-knight.

Keywords: **chess, expert system, inference, knowledge acquisition**

F51 Reboh, R.
Knowledge Engineering Techniques and Tools for Expert Systems
Linköping Studies in Science and Technology Dissertation No. 71 (1981)

This report explores techniques and tools to assist in several phases of the knowledge-engineering process for developing an expert system. A sophisticated domain-independent network editor is described that uses knowledge about the representational and computational formalisms of the host consultation system to watch over the knowledge-engineering process and to give the technology engineer a convenient environment for developing, debugging and maintaining the knowledge base. The ways in which partial matching techniques can assist in maintaining the consistency of the knowledge base (in form and content) as it grows, and can support a variety of features that will enhance the interaction between the system and the user and make a knowledge-based consultation system behave more intelligently are illustrated. The techniques and features are illustrated in terms of the Prospector environment but it is suggested that they can be applied in other environments.

Keywords: **consultation system, expert system, knowledge base, knowledge engineering, man-machine interface, Prospector**

F52 Roberts, S. K.
Artificial Intelligence
Byte, Vol. 6, No. 9, pp.164–178 (September 1981)

If computers are to come to terms with heuristic as opposed to merely factual knowledge, they must possess a measure of intelligence. This paper traces the origins and current status of the discipline of artificial intelligence and considers the problems that must be overcome if computers are to progress towards the goal of being intelligence amplifiers. The problems posed by knowledge representation, natural language and the human brain's ability to perform many operations in parallel are analysed.

Keywords: **artificial intelligence, human knowledge processing, knowledge representation, natural language, parallel processing**

F53 Rychener, M. D.
Knowledge-Based Expert Systems: A Brief Bibliography
Carnegie-Mellon University, Department of Computer Science Technical Report No. CMU-CS-81-127

This paper gives a selective reading list on the subject of knowledge-based

Expert Systems. The entries are intended not to be excessively technical and usually contain references to related work and background material. There are a total of 42 items included.

Keywords: **bibliography, expert system**

F54   Sacerdoti, E. D.
Practical Machine Intelligence
In (F32), pp.241–247

Machine Intelligence Corporation (MIC) was founded in 1978 as a vehicle for bringing the more practical aspects of the machine intelligence field into widespread use. Three specific areas that MIC feels are ripe for commercial application are discussed: machine vision, natural language access to computers and Expert Systems.

Keywords: **expert system, Machine Intelligence Corporation, natural language, vision**

F55   Sato, S.
Fujitsu Research Activities in Knowledge Information Processing Systems
In (A3)

A number of research activities in knowledge information processing systems at the Fujitsu company are described including the LISP machine, Alpha; the machine translation systems, ATLAS I and ATLAS II; a pattern recognition system for Chinese characters; image processing systems for recognising hand drawn wiring and circuit diagrams, and for recognising three dimensional objects as the basis for a fault recognition system; developments in speech recognition; developments in intelligent CAD; developments concerned with a relational data base system. The paper concludes by identifying inference and problem solving techniques as areas for possible future research activity. Printed copies of the transparencies used by the author in his conference presentation are appended to the paper.

Keywords: **Alpha, Atlas I and II, CAD, Fujitsu, image processing, inference, Japan, Lisp machine, machine translation, pattern recognition, problem solving, relational database, voice recognition**

F56   Scarrott, G.
The Need for Theory to Illuminate Problem Solving and Inference
In (A12), pp.115–122

This paper discusses the need for and the function of theory in engineering and information engineering and stresses the importance of using theoretical understanding to design machines that will meet human needs. The concept of information is a familiar one, but although information is used in every aspect of human life, it is not easy to define its role in human affairs. The vital role of information is considered, with particular reference to the main aspects of information engineering — storage retrieval and communication of symbols — and the processing of these symbols to solve current problems. The need to create a valid theoretical model for the latter aspects of information engineering is emphasised. Finally, possible conceptual foundations for a theory of information are discussed.

Keywords: **inference, information technology, problem solving, social context**

F57   Sell, P. S.
New Computer Applications: User and Social Acceptability of the Fifth Generation Proposals
In (A2)

The Japanese plans for a Fifth Generation computer, if successfully carried out, will offer a host of new facilities and capabilities to their users. This paper examines some of the more important effects of these facilities and their impact. It considers separately the impact of the three major changes advocated, namely, the emphasis on problem solving, logical inference, and use in areas of judgement.

Keywords: **expert system, FGCS project, inference, Japan, problem solving**

F58   Sparck-Jones, K.
UK Initiatives on Intelligent Knowledge-Based Systems
AISB Quarterly, No. 45, pp.23-25 (Autumn 1982)

This article describes five important United Kingdom initiatives in intelligent knowledge based systems (IKBS) during 1982: 1) the Science and Engineering Research Council (SERC) proposal for a Specially Promoted Programme in IKBS; 2) the working paper of the Department of Industry (DoI) committee set up to make recommendations for activity in information technology; 3) the Alvey Committee Report, A Programme for Advanced Information Technology; 4) the proceedings of the SERC research area review meeting on IKBS held in London in September; 5) a project funded by the DoI and the SERC to carry out the architecture study advocated by SERC.

Keywords: **Alvey programme, architecture, IKBS, information technology, United Kingdom**

F59   Stefik, M., Aikins, J., Balzer, R., Benoit, J., Birnbaum, L., Hayes-Roth, F. and Sacerdoti, E.
The Organization of Expert Systems: A Prescriptive Tutorial
In (A2)
Also available as Xerox Palo Alto Research Center Research Report VLSI-82-1 (January 1982)

In August 1980 the National Science Foundation and the Defense Advanced Research Projects Agency co-sponsored a workshop on Expert Systems in San Diego. The participants were organised in groups to produce materials suitable for the chapters of a definitive book on Expert Systems. This paper is the product of the 'architecture' group. The tutorial begins with a brief review of standard topics from artificial intelligence, such as the use of the predicate calculus and search techniques. It also discusses some newer topics which are essential to an understanding of current work, such as the use of abstractions, assumptions, dependent information, and meta-level cognition. The main section of the tutorial is a discussion of organisational prescriptions for problem solvers. It begins with a restricted class of problem that admits a very simple organisation. The requirements are then relaxed, one at a time, yielding 10 case studies of organisational prescriptions. The first cases describe techniques for dealing with unreliable data and time-varying data. Other cases illustrate techniques for creating abstract solution spaces and using multiple lines for reasoning. The prescriptions are compared for their coverage and illustrated by examples from recent Expert Systems.

Keywords: **artificial intelligence, expert system, search**

F60 Suwa, M., Furukawa, K., Makinouchi, A., Mizoguchi, T., Mizoguchi, F. and Yamasaki, H.
Knowledge Base Mechanisms
In (A11), pp.139–145
This paper describes the Japanese plan of R & D on knowledge-base mechanisms as part of the Fifth Generation Project. It is aiming to design a knowledge representation system to support knowledge acquisition for knowledge information processing systems. The system includes a knowledge representation language, a knowledge-base editor and a debugger. It is also expected to perform as a kind of meta-inference system. With respect to large-scale knowledge-base systems, a knowledge-base mechanism based on the relational model is to be studied in the early stage of the project. Distributed problem solving is also a main issue of the work.

Keywords: **distributed processing, FGCS project, Japan, knowledge acquisition, knowledge base, knowledge engineering, knowledge representation**

F61 Vince, N. L.
Decision Support Meets Expert Systems
In (A3)

Transcript of the talk presented at the conference, together with printed copies of the transparencies used. The objective of the talk was to illustrate how the scope and power of decision support systems could be enhanced by the better use of knowledge, and knowledge engineering technology, particularly with expert systems. This is done through an examination of the main categories of decision making and the nature of the decision process — at the operational, managerial and policy levels — including stages of intelligence gathering, option articulation, modelling, choice and commitment.

Keywords: **decision support system, expert system, knowledge engineering**

F62 Webster, R.
Planting an Expert
Micro Decisions, No. 14, pp.107-108 (December 1982)

The author argues that expert systems on microcomputers are now feasible. He explains how an expert system can solve a problem and explains its methods. In particular he examines a plant care expert system which is a demonstration program for a micro-based expert system builder called Micro-Expert from Isis Systems.

Keywords: **expert system, expert system shell, Isis Systems, Micro-expert, microcomputer, plant care**

F63 Webster, R. and Miner, L.
Expert Systems: Programming Problem-Solving
Technology 2, pp.62-73 (January/February 1982)

This paper examines the nature of Expert Systems, considers how they are built, describes inter alia three major systems — DENDRAL, MYCIN, and PROSPECTOR — discusses the extent of business interest in Expert Systems and presents information about Teknowledge, Inc, a company that provides training courses and software related to Expert Systems. A list of research centres active in the field and of relevant publications is appended.

Keywords: **Dendral, expert system, Mycin, Prospector, Teknowledge Inc**

F64 Welbank, M.
A Review of Knowledge Acquisition Techniques for Expert Systems
Martlesham Consultancy Services (1983)

Few reports of expert system projects describe the knowledge acquisition process, but it is frequently cited as the major bottleneck to building an expert system. This report has been produced as part of an investigation of the human factors of expert systems, currently in progress at the British Telecom Research Laboratories. It is based on a review of the literature and on experiences gained from a number of exercises in knowledge acquisition undertaken within British Telecom. The aim is to provide information that would be practically useful to anyone taking on an expert system project and seeks to indicate how knowledge acquisition is done, and how and why it is difficult. The problems that can be encountered during knowledge acquisition are discussed under the three stages of: 1) initially structuring the domain; 2) producing the first working system, and 3) testing and debugging the system. A discussion of machine induction concludes the report.

Keywords: **automatic induction, British Telecom Research Laboratories, expert system, knowledge acquisition**

F65 Winfield, M. J.
Expert Systems: An Introduction for the Layman
Computer Bulletin, II/34, pp.6–7, 18 (December 1982)

This article begins by offering a definition of an expert system and describing its essential parts, namely a knowledge base, an inference engine (driver program), a natural language front end translator program, an explanatory capability and a program for updating the knowledge base. It goes on to consider the range of problem domains where expert systems might be used and to justify the use of an expert system to solve a particular problem.

Keywords: **expert system, explanatory capability, inference engine, knowledge base, natural language**

F66 Woods, W. A.
What's Important About Knowledge Representation?
Computer, Vol. 16, No. 10, pp.22–27 (October 1983)

This article explores the question of what constitutes a good representational system and a good set of representational primitives for dealing with an open-ended range of knowledge domains. Two aspects of knowledge representation are examined, the role of knowledge networks in intelligent machinery and several epistemological considerations for knowledge structures. The techniques and concepts evolved while developing the knowledge-representation system KL-One are used as illustrations. The article concludes by arguing the need for taxonomic organisation as a means towards advancing expressive adequacy and notational efficacy for intelligent systems.

Keywords: **KL-One, knowledge representation**

F67 Zadeh, L. A.
The Role of Fuzzy Logic in the Management of Uncertainty in Expert Systems
Fuzzy Sets and Systems, Vol.11, No. 3, pp.199–227 (November 1983)

Management of uncertainty is an intrinsically important issue in the design of expert systems because much of the information in the knowledge base of a typical expert system is imprecise, incomplete, or not totally reliable. In the existing expert systems, uncertainty is dealt with through a combination of predicate logic and probability-based methods. A serious shortcoming of these methods is that they are not capable of coming to grips with the pervasive fuzziness of information in the knowledge base and, as a result, are mostly ad hoc in nature. An alternative approach to the management of uncertainty which is suggested in this paper is based on the use of fuzzy logic, which is the logic underlying approximate or, equivalently, fuzzy reasoning. A feature of fuzzy logic which is of particular importance to the management of uncertainty in expert systems is that it provides a systematic framework for dealing with fuzzy quantifiers, e.g. most, many, few, not very many, almost all, infrequently, about 0.8, etc. In this way, fuzzy logic subsumes both predicate logic and probability theory, and makes it possible to deal with different types of uncertainty within a single conceptual framework. In fuzzy logic, the deduction of a conclusion from a set of premises is reduced, in general, to the solution of a nonlinear program through the application of projection and extension principles. This approach to deduction leads to various basic syllogisms which may be used as rules of combination of evidence in expert systems. Among syllogisms of this type which are discussed in this paper are the intersection/product syllogism, the generalised modus ponens, the consequent conjunction syllogism, and the major-premise reversibility rule.

Keywords: expert system, fuzzy logic, plausible reasoning

G | Networks

Abstract Numbers: G1-G3

G1 Gottlieb, A. and Schwartz, J. T.
Networks and Algorithms for Very-Large-Scale Parallel Computation
Computer, Vol. 15, No. 1, pp.27-36 (January 1982)

This paper describes the ultracomputer — a computer constructed from large numbers of standard microprocessor chips that are tightly coupled in a suitable network.

Keywords: **networks, parallel processing, ultracomputer, VLSI**

G2 Kobayashi, K.
Computer, Communications and Man: The Integration of Computer and Communications with Man as an Axis
Computer Networks, Vol. 5, No. 4, pp.237-250 (July 1981)

In this paper, the President of the Information Processing Society of Japan argues the importance of integrating computers and communications in the coming era (from the 1980s to the year 2000). In this context human factors are of paramount interest. It is extremely difficult to express the direction of development of computers and communications merely by means of the two-dimensional x-y plane (with 'computers' and 'communications' as the two axes). It is necessary to use the three-dimensional x-y-z plane, with 'man' as the third axis. The important role of software in the creation of a 'man, computers and communications' society is also considered.

Keywords: **networks, social context**

G3 Pouzin, L.
Networks for the Fifth Generation
In (A2)

At first sight, the Japanese Fifth Generation Computer Project is paving the way towards a world of interconnected computers handling and augmenting human knowledge. Without questioning this goal, the article discusses some technical implications of distributed systems based on characteristics stated by the Japanese and those inferred from projections into a futuristic age. Characteristics of the former kind include bit rate, transit delay, integrity and availability and those of the latter kind include standardisation, networking facilities, user assistance and protection. Major predictable roadblocks are finally considered. Printed copies of the transparencies used by the author in his conference presentation are appended to the paper.

Keywords: **distributed processing, networks**

Reference Sources

The following list is designed to enable any of the documents included in the bibliography to be obtained as easily as possible.

It includes every journal, magazine, etc. cited, with the name and address of its publisher, plus the name and address of the publisher of every book included in the bibliography and addresses of sources from which the remaining items (such as technical reports), which usually have a more restricted circulation than those in books and periodicals, can be obtained.

Academic Press Inc (London) Ltd
24–28 Oval Road
London NW1 7DX
UK

Addison-Wesley Publishers Ltd
Finchampstead Road
Wokingham
Berkshire RG11 2NZ
UK

and

Addison-Wesley Publishing Co Inc
Reading
Massachusetts 01867
USA

AI Magazine
American Association for Artificial Intelligence
445 Burgess Drive
Menlo Park
California 94025
USA

AISB Quarterly
Society for the Study of Artificial Intelligence and Simulation of Behaviour
Secretary: Dr. R. Young
MRC Applied Psychology Unit
Cambridge
UK

Alvey Directorate
Department of Trade and Industry
Millbank Tower
Millbank
London SW1P 4QU
UK

Aslib Proceedings
Aslib
3 Belgrave Square
London SW1X 8PL
UK

British Computer Society Specialist Group on Expert Systems
c/o Imperial Cancer Research Fund
Lincoln's Inn Fields
London WC2A 3PX
UK

Brunel University Man-Computer Studies Group
Kingston Lane
Uxbridge
Middlesex UB8 3PH
UK

Business Week
McGraw-Hill Inc
1221 Avenue of the Americas
New York
NY 10020
USA

Byte
Byte Publications Inc
70 Main Street
Peterborough
New Hampshire 03458
USA

Cambridge University Press
The Edinburgh Building
Shaftesbury Road
Cambridge CB2 2RU
UK

Carnegie-Mellon University
Department of Computer Science
Schenley Park
Pittsburgh
Pennsylvania 15213
USA

Communications of the ACM
Association for Computing Machinery
1133 Avenue of the Americas
New York
NY 10036
USA

Computer
IEEE Computer Society
10662 Los Vaqueros Circle
Los Alamitos
California 90720
USA

Computer Bulletin
John Wiley & Sons Ltd
Baffins Lane
Chichester
West Sussex PO19 1UD
UK

Computer Journal
(As for Computer Bulletin)

Computer Networks
North-Holland Publishing Company
PO Box 103
1000 AC Amsterdam
The Netherlands

Computerworld
CW Communications Inc
375 Cochituare Road, Box 880
Farrington
Massachusetts 01701
USA

Computing Europe
VNU Business Publications Ltd
53–55 Frith Street
London W1A 2HG
UK

Computing Surveys
(As for Communications of the ACM)

Datamation
Technical Publishing Company
666 Fifth Avenue
New York
NY 90035
USA

DecNews
Digital Equipment Corporation
129 Parker Street
Maynard
Massachusetts 01754
USA

Edinburgh University Press
22 George Square
Edinburgh
UK

Electronics
(As for Business Week)

Ellis Horwood Ltd
Market Cross House
Cooper Street
Chichester
West Sussex PO19 1EB
UK

Fuzzy Sets and Systems
Elsevier Science Publishers BV
(North-Holland)
PO Box 211
1000 AE Amsterdam
The Netherlands

Gordon and Breach Science
Publishers Ltd
42 William IV Street
London WC2N 4DE
UK

HMSO
Her Majesty's Stationery Office
49 High Holborn
London WC1V 6HB
UK

ICOT
Institute for New Generation
Computer Technology
Mita Kokusai Building
21st Floor 1-4-28
Mita
Minato-ku
Tokyo 108
Japan

ICOT Journal
(As for ICOT)

IEEE
Institute of Electrical and Electronics
Engineers, Inc
345 East 47th Street
New York
NY 10017
USA

IEEE Spectrum
(As for IEEE)

Imperial College of Science
and Technology
Department of Computing
180 Queens Gate
London SW7 2BZ
UK

Informatics
(As for Computing Europe)

Institution of Electrical Engineers
Savoy Place
London WC2R 0BL
UK

International Journal of Man-Machine
Studies
(See Academic Press Inc (London) Ltd)

JIPDEC
Japan Information Processing
Development Center
Kikai Shinko Kaikan Building
5-8 Shibakoen 3-chome
Minato-ku
Tokyo 105
Japan

Linköping Studies in Science and
Technology — Dissertations
Linköping University
S-58183 Linköping
Sweden

Martlesham Consultancy Services
TEL 1.2
British Telecom Research Laboratories
Ipswich
Suffolk IP5 7RE
UK

Microdecision
(As for Computing Europe)

Nature
Macmillan Journals Ltd
Brunel Road
Basingstoke
Hampshire RG21 2XS
UK

NCC Publications
The National Computing Centre Ltd
Oxford Road
Manchester M1 7ED
UK

New Generation Computing
Springer-Verlag
37a Church Road Wimbledon
London SW19
UK

and

175 Fifth Avenue
New York
NY 10010
USA

and

D-1000 Berlin 33
Heidelbergerplatz 3
West Germany

North-Holland Publishing Company
(See Computer Networks)

Pergamon Infotech Ltd
Berkshire House
Queen Street
Maidenhead
Berkshire SL6 1NF
UK

Pitman Books
128 Long Acre
London WC2E 9AN
UK

Prentice-Hall International
66 Wood Lane End
Hemel Hempstead
Hertfordshire HP2 4RG
UK

Schlumberger-Doll Research
Ridgefield
Connecticut
USA

Science
American Association for the
Advancement of Science
1515 Massachusetts Avenue NW
Washington
DC 20005
USA

Science and Engineering Research
Council (SERC)
Polaris House
North Star Avenue
Swindon
Wiltshire SN2 1ET
UK

Scientific American
Scientific American Inc
415 Madison Avenue
New York
NY 10017
USA

SERC Rutherford Appleton Laboratory
Chilton
Didcot
Oxfordshire OX11 0QX
UK
(Attn: Mr W. Sharp)

SIGART Newsletter
ACM Order Department
PO Box 64145
Baltimore
Maryland 21264
USA

SPL International
SPL International Research Centre
The Charter
Abingdon
Oxfordshire OX14 3LZ
UK
(Attn: Ms. W Allen)

Springer-Verlag
(As for New Generation Computing)

Stanford Heuristic Programming
Project
Computer Science Department
Stanford University
Stanford
California 94305
USA

Techniques et Sciences Informatiques
Cie. Electro-Mechanique
37 Rue du Rocher
Paris 8
France

Technology
Technology Information Corporation
Colorado
USA

Tioga Publishing Co.
PO Box 98
Palo Alto
California 94302
USA

University of Cambridge Computer
Laboratory
Corn Exchange Street
Cambridge CB2 3QG
UK

University of Newcastle-upon-Tyne
Computing Laboratory
Claremont Tower
Newcastle-upon-Tyne NE1 7RU
UK

VLSI Design
Redwood Systems Group
PO Box 50518
Palo Alto
California 94303-0518
USA

Wiley Heyden Ltd
(As for Computer Bulletin)

Xerox Palo Alto Research Center
3333 Coyote Hill Road
Palo Alto
California 94304
USA

Author Index

Ackerman, W. B., D1
Addis, T. R., F6, F7
Aida, H., E1
Aikins, J., F59
Aiso, H., D2, D33
Allen, J., C1; D3, D4
Amamiya, M., D5, D39
Aoyama, A., F27
Arvind, D6

Backus, J., E2
Ballard, D. H., C2
Balzer, R., B6; F59
Barnett, J. A., F8
Barr, A., F9
Basden, A., F10
Bell, A., D38
Bell, D. H., D7, D8
Bell, T. E., B5
Benoit, J., F59
Bibel, W., F11
Bigger, C. J., F12
Bird, J., B7; F13
Birnbaum, L., F59
Bloch, E., B5
Bobrow, D. G., D38
Boley, H., B8
Boral, H., D9
Bott, M. F., C3
Bramer, M. A., F14
Brandin, D. H., B9
Brown, H., D38
Brownbridge, D. R., D46
Bruckert, E., C4
Buchanan, B. G., F15, F16
Bundy, A., F17, F18

Carbonell, J. G., F42
Cercone, N., F39
Chandrasekaran, B., F19
Cheatham, T. E. Jr., B6
Chiba, S., C17
Chikayama, T., E3
Clark, K. L., E4, E5, E6

Clocksin, W. F., E7
Cohen, C., D10
Connolly, R., B10
Conway, L., D28, D38
Cooper, R. S., B5
Coupland, J. W., F12

d'Agapeyeff, A., B11, B12; F20
Darlington, J., E8
Davis, A. L., B5; D11
De Jong, K. A., F21
Dennis, J. B., D12
Dewitt, D. J., D9
Douglass, R. J., B5
Duda, R. O., F22, F23
Duff, M. J. B., C5
Dusek, L., C6
Dyer, C. R., C13

Elcock, E. W., F24
Erman, L. D., C7
Evanczuk, S., C8; F40
Evans, D. J., D13

Fairbairn, D. G., D14
Farr, R. B., B5
Feigenbaum, E. A., B5, B13, B14; F9, F25, F26
Ferguson, R., E9
Ferris, D., C12
Fischetti, M. A., B5
Fuchi, K., B15, B16
Funifuji, S., F36
Furukawa, K., E10, E23, E33; F27, F36, F60
Fusaoka, A., D5

Gajski, D. D., D15
Galinski, C., D16
Gannon, T. F., B17
Garvey, T. D., F38
Gaschnig, J. G., F22
Gevarter, W. B., F28, F29
Gostelow, K. P., D6
Goto, A., D39

Goto, S., E11, E37; F27
Gottlieb, A., G1
Green, C., B6
Gregory, S., E12
Gurd, J., D17, D53
Guterl, F., B5

Hakozaki, K., D5
Hardy, S., E31
Harkness, D. L., D18
Harris, L.C9; F30
Hattori, T., E13
Hawkins, D., F31
Hayashi, H., E37
Hayes, J. E., F32
Hayes-Roth, F., B5; C7; F33, F59
Haynes, L. S., D19
Haynes, S., C11
Henderson, P., E8
Hendrix, G., C10
Hikita, T., E14
Hinton, G. E., C2
Hirakawa, H., E15, E16
Hirsch, P. M., F34
Hollander, C. R., F34
Hopkins, R. P., D46, D47
Horstmann, P. W., D20
Hudson, K., B18
Hunt, D. J., D21

Ito, N., D22, D39

Jain, R., C11
Johnson, J., B19
Johnson, L., F7
Jordan, J. A. Jr., F34

Kadowaki, Y., D39
Kahn, G., F35
Kahn, R. E., B5
Kakuta, T., D23, D30, D34, D35
Kanade, T., B5
Kaplan, S. J., C12
Karatsu, H., B20
Kawatani, Y., B21
Keller, R. M., D11
Kent, E., C13
Kerridge, J. M., D7
Keyworth, G. A. II, B5
Kidode, M., C17
Kim, K. H., B22
King, R., B5

Kitakami, H., E23; F36
Kitsuregawa, M., D24, D39
Kobayashi, K., G2
Kodaka, T., D33
Kodera, T., C17
Kowalski, R., B23; E17, E18, E19, E20
Kuck, D. L., D15
Kuhn, R. H., D15
Kung, H. T., D25
Kunifufi, S., E23
Kunifuji, S., E21, E37; F36
Kurokawa, T., E37

Langlotz, C. P., F37
Lau, R. L., D19
Lehman, M. M., B24
Lemmons, P., B25
Lenat, D. B., F33
Lesser, V. R., C7
Lewicki, G., D29
Lima, I. G., B40
Lowrance, J. D., F38
Lumley, J., F17

McCabe, F. G., E5
McCalla, G., F39
McCorduck, P.B14, B30
McCormick, B. H., C13
McMahan, M., C6
Makinouchi, A., F60
Malik, R., B26, B27
Maller, V. A. J., D26
Manuel, T., B28, B29; C8; D27; E22; F40
Masuda, K., D22
Mead, C. A., D28, D29
Meindl, J. D., B5
Mellish, C. S., E7
Merry, M., F17
Michaelsen, R., F41
Michalski, R. S., F42
Michie, D., B31; F32, F41, F43, F44, F45
Mill, J., B32
Mills, A. F., D18
Miner, L., F63
Minow, M., C4
Mitchell, T. M., F42
Mittal, S., F19
Miyachi, T., E23; F36
Miyazaki, N., D23, D30, D34, D35
Mizell, D. W., D19
Mizoguchi, F., F46, F60

Mizoguchi, T., F60
Moto-oka, T., A10, A11; B5, B33; C14; D24; E1
Motoyoshi, F., E37
Murakami, K., D23, D30, D35
Myers, W., E24

Naisbitt, J., B5
Nakajima, R., E10; F27
Nakashima, H., E25, E37
Nasko, H., B5
Nau, D. S., F47
Neff, R., C15, C16
Nishikawa, H., D31, D52
Nitta, K., E37

Oakley, B. W., B5
Onai, R., D22

Padua, D. A., D15
Pao, Y-H., F32
Pease, R. F. W., B5
Plummer, D., F18
Pouzin, L., G3

Quinlan, J. R., F48, F49, F50

Raj Reddy, D., B5; C7
Randell, B., D32
Rautenbach, P. W., D47
Reboh, R., F51
Reddaway, S. F., D21
Reinstein, H. C., F34
Roberts, S. K., F52
Robinson, J. A., E26, E27
Rychener, M. D., F53

Sacerdoti, E., C10; F54, F59
Sakamura, K., D33
Sakura, T., E28
Sántáné-Tóth, E., E32
Sato, M., E28
Sato, S., E34; F55
Sato, T., E35, E37
Scarrott, G. G., A12; F56
Schalk, T. B., C6
Schwartz, J. T., G1
Sejnowski, T. J., C2
Sekino, A., D33
Sell, P. S., F57
Shapiro, E., B34; E29, E30
Sheil, B., B5
Shibayama, S., D23, D30, D34, D35

Shimada, T., D39
Shimizu, H., D22
Shiraishi, T., E37
Shortliffe, E. H., F23, F37
Sibert, E. E., E27
Siewiorek, D. P., D19
Silver, B., F18
Simons, G. L., B35
Simpson, D., D7, D8
Sloman, A., E31
Smith, K., D36
Snyder, L., D37
Sohma, Y., D39
Sparck-Jones, K., A13; F17, F58
Spennewyn, D., B36
Stefik, M., D38; F59
Steier, R., B37
Sugimoto, M., D51
Sumner, F. H., B38
Suwa, M., F60
Suzuki, N., E25
Szeredi, P., E32

Takei, K., D51
Taki, K., D31, D52
Takeuchi, A., D39; E30, E33
Takeuchi, S., E23
Takizawa, M., D39
Tamaki, H., E34, E35
Tamura, H., C17
Tanaka, H., C17; D24, D39, D51; E1
Tanaka, K., B39
Tanaka, Y., D5, D39
Tarnlund, S. A., E6
Tetschner, W., C4
Tokoro, M., D51
Tong, C., D38
Torrero, E. A., B5
Treleaven, P. C., B40; D32, D40, D41, D42, D43, D44, D45, D46, D47
Trimberger, S., B5
Turner, D. A., D48; E8

Uchida, S., D31, D49, D50, D51, D52
Ueda, K., E37
Uehara, T., D33
Umemura, M., E37
Umeyama, S., E37
Underwood, M. J., C18

Vince, N. L., F61

Wallich, P., B5
Waltz, D. L., B5, B41
Warren, D. H. D., B42; E36
Waterman, D. A., F33
Watson, I., D53
Webster, R., F62, F63
Weil, U., B43
Welbank, M., F64
Werner. J., D54, D55
Willis, N., D7
Winfield, M J., F65
Withington, F. G., B44
Woods, W. A., F66

Yamamoto, A., D31, D52
Yamamoto, M., D39
Yamasaki, H., F60
Yasaki, E. K., B45
Yasuhara, H., D51
Yasukawa, H., C19
Yeh, R., B5
Yianilos, P. N., D56
Yokoi, T., D5; E13, E37
Yokota, H., D23, D30, D34, D35; E21, E23
Yokota, M., D31, D52
Yonezawa, A., E10; F27

Zadeh, L. A., F67
Zue, V., B5

Subject Index

Absys, F24
ACM (Association for Computing Machinery), B9, B37
Adips, F46
Age, F2
ALICE (Applicative Language Idealised Computing Engine), D36; E22
Alpha, F55
Alvey programme, A1, A4, A8, A13; B1, B13, B18, B32, B35; F58
AL/X, F49
AM, F26
Aplicot, F46
architecture, A6, A10; B4, B5, B15, B17, B28, B31, B35, B38, B39, B45; C13; D2, D3, D5, D9, D12, D17, D22, D24, D25, D27, D29, D31, D32, D40, D41, D42, D44, D45, D46, D49, D50, D51; E8, E22, E29; F2, F11, F28, F48, F58
(See also dataflow, graph reduction, parallel processing)
array processor, B36
(See also Distributed Array Processor)
artificial intelligence, B3, B5, B8, B15, B20, B22, B23, B32, B35, B38, B39, B41; C8, C12; D19, D38; E22, E24, E31; F1, F2, F3, F5, F9, F17, F23, F32, F40, F42, F43, F47, F52, F59
Atlas I and II, F55
Australia, B3
automatic induction, F42, F43, F64
automatic programming, B31, B33
automatic test generation, D20

Bacon, F42
Bayesian system, F49
Belgium, B5
bibliography, A12; B5; D7; F12, F14, F16, F33, F42, F53
blackboard model, F2, F48
British Telecom Research Laboratories, F64
business application, B12; C15; F41

CAD (Computer Aided Design), A1; B4, B17, B33; D2, D20, D33, D42; E32; F55
CAI (Computer Aided Instruction), F35
CAM (Computer Aided Manufacture), B4, B17, B21
catalogue of AI tools, F17
chess, F42, F45, F50
CLIP4 (Cellular Logic Image Processor), C5
collection of papers, A2, A3, A9, A11, A12; B5; D13, D42; E6; F4, F32, F39, F42, F44, F45
Concurrent Prolog, E15, E16, E29, E30, E33
Congen, F44
consultation system, B33; F37, F51
Content Addressable File Store, D26
control flow, D27, D40, D41, D47
control problem, F8
Cray computer, B36
Crysalis, F26
Cyber 205 computer, B36

Dart, F2, F34
data driven computer, D46
database, B20, B45; C8; D56
(See also relational database)
database machine, D24, D30, D34, D35, D39
database query system, C12; E17, E20
dataflow, B3, B7, B35, B45; D1, D6, D9, D11, D12, D15, D17, D18, D19, D22, D27, D39, D40, D41, D46, D47, D48, D49, D51, D53; E29
decision support system, F41, F61
Delta, D23, D30, D34, D35
demand driven computer, D46
Dendral, C8; F26, F28, F63
Denmark, B5
Digital Equipment Corporation, C4; F3
Direct, D9
Distributed Array Processor, D21
distributed processing, A10; B36, B40;

D21, D40, D41; F60; G3
document preparation, F35
education and training implications, A7; B7
educational application, F5, F45
Emycin, F46
engineering application, F5
ESP (Extended Self-contained Prolog), E3
Esprit programme, B5, B32
ETL (Electrotechnical Laboratory), B22, B23
Eurisko, F2
Eurotra project, C16
evidential reasoning, F38
EXES/370, F34
Expert, F46
expert system, A7; B5, B7, B18, B23, B31, B35, B36, B38, B40, B45; C3, C8; D20, D40; E4, E5, E17, E18, E20, E24; F1, F2, F3, F4, F5, F6, F8, F10, F12, F13, F14, F15, F16, F19, F20, F21, F22, F23, F28, F29, F30, F31, F33, F34, F37, F38, F40, F41, F44, F45, F46, F47, F48, F49, F50, F51, F53, F54, F57, F59, F61, F62, F63, F64, F65, F67
(See also consultation system, expert system shell, IKBS, and specific systems, e.g. Mycin)
expert system shell, C8; E5; F4, F5, F46, F62
expert system teaching aid, F22
explanatory capability, E5; F15, F26, F37, F65
Export Software International, B32
extended relational analysis, F6

fault finding, D20; E5; F46
FGCS project (Japanese Fifth Generation Computer Systems), A5, A6, A9, A10, A11, A12; B2, B5, B9, B11, B13, B14, B16, B18, B19, B22, B23, B24, B25, B26, B27, B28, B29, B30, B33, B34, B37, B38, B39, B40, B42, B45; C14, C18, C19; D10, D30, D31, D34, D35, D39, D40, D49, D50, D51, D52; E3, E10, E37; F20, F27, F30, F43, F57, F60
financial application, E5
frame, F7, F39, F48
France, B5; C8; D27
Formel, F35

Fujitsu, B21, B22; F55
functional programming, C8; D1, D18, D19, D42, D48; E2, E8, E14, E20
fuzzy logic, F67
fuzzy sets, C11

geological application, F31
Glisp, F2
Grace, D24
graph reduction, B7; E14, E16
Greece, B5
Guidon, F2

Hearsay II, C7
Hitachi, B22
Hope, E22
human knowledge processing, B18, B41; C2, C13; F31, F52
Hungary, E32

IA (Intelligent Agent), F2
ICL (International Computers Ltd.), B7; D21, D26
ICOT (Institute for New Generation Computer Technology), B23, B30, B34; C19; D10, D30, D35; E3
IKBS (Intelligent Knowledge Based System), A4, A7, A13; B1; F58
(See also expert system)
image processing, B45; C5, C14, C17; D21; F55
Imperial Cancer Research Fund, B32
Imperial College, B7; D36; E22
incomplete knowledge, F24
inference, A6, A7; B15, B28, B33; D5, D51; E22; F27, F36, F49, F50, F55, F56, F57
(See also reasoning)
inference engine, F48, F65
inference machine, D49, D50, D51; E6; F30
Inferno, F49
information retrieval, E32; F6
information technology, B38; F56, F58
Information Technology Promotion Agency, B22
Inmos, B7; D32, D36
integrated circuit technology, B35
intelligent interface, A6; B28, B30, B33; C14, C17, C18
international relations, A5, A10; B2, B3,

B4, B20, B33, B44; D16; F25
Internist, F28
Ireland, B5
Isis Systems, B7; F62
Italy, B5

Japan, A5, A6, A9, A10, A11, A12; B2, B3, B5, B9, B11, B13, B14, B16, B18, B19, B20, B21, B22, B23, B25, B26, B27, B28, B29, B30, B33, B34, B38, B39, B40, B42, B43, B45; C14, C15, C16, C17, C18, C19; D10, D16, D27, D30, D31, D34, D35, D39, D40, D46, D49, D50, D51, D52, D54; E3, E22, E37; F20, F30, F40, F43, F55, F57, F60

Keio University, B22
KL0, D31, D52; E3, E13
KL-One, F66
knowledge acquisition, E23; F2, F4, F7, F15, F18, F36, F39, F45, F48, F50, F60, F64
knowledge base, A6; B28, B33; D5; F26, F30, F47, F51, F60, F65
knowledge base machine, D30, D34
knowledge based system, B12, B44; C8; F25, F32, F36
knowledge engineering, A10; B7, B15, B16, B39, B45; D38, D51; E10; F5, F7, F21, F26, F45, F51, F60, F61
knowledge representation, A7; B5, B15; F7, F9, F15, F19, F23, F24, F39, F47, F48, F52, F60, F66
KRC, D48
Kyoto University, B23

Ladder, C10
LCF, F35
LFG (Lexical Functional Grammar), C19
Lisp, C8; E22, E24, E28, E31
(See also Glisp, Lisp machine, Loglisp)
Lisp machine, E6; F55
logic machine, B23, B42
logic programming, B7, B16, B34, B35; D2, D31, D33, D34, D49, D51, D52; E3, E4, E5, E6, E7, E10, E11, E12, E15, E17, E19, E20, E21, E23, E26, E27, E28, E30, E34, E36; F32, F45
(See also specific programming languages, e.g. Prolog)
Logic Programming Associates, E22
Loglisp, E27

LOPS (Logical Program Synthesis System), F11
LSI (Large Scale Integration), B44
(See also VLSI)
Loughborough University, D13

Machine Intelligence Corporation, F54
machine learning, B28; F2, F18, F42, F45
machine translation, B33, B45; C3, C8, C16, C17; F55
Macsyma, F35
man-machine interface, A7, A10; B1, B20; F51
(See also intelligent interface)
Manchester University, B7
mathematical application, D21
MCC (Microelectronics and Computer Technology Corporation), B4, B17, B37; D55
MCE Inc, B10
MDX, F19
Mecho, F44
medical application, E5, E32; F5, F12, F19
Mentor, F35
Meta-Dendral, F26
Micro-expert, F62
Micro-Prolog, E5
microcomputer, D42; F62
(See also personal computer)
MIRA (Machine Intelligence Research Affiliates), B31
MITI (Ministry of International Trade and Industry), B2, B3
Molgen, F2, F26
MRS (Metalevel Representation System), F2
Mycin, F23, F26, F28, F63
(See also Emycin, Neomycin)

natural language, A7; B5, B15, B23, B41, B45; C8, C9, C10, C12, C14, C17, C19; E6, E24; F9, F30, F40, F52, F54, F65
Navy Centre for Applied Research in AI, F21
NEC, B23
Neomycin, F2/11
Netherlands, B5
networks, B5, B20, B28; D33; G1, G2, G3
Nippon Electric Company, B22
NTT (Nippon Telephone and Telecommunications), B23

object-oriented programming, C8; E25,

E30
Oncocin, F2, F37

Palladio, D38
parallel processing, B3, B27; C2, C5; D1, D6, D7, D8, D9, D12, D13, D14, D17, D19, D21, D22, D25, D27, D36, D37, D42, D48, D49, D50, D51, D56; E12, E15, E22, E29, E30, E33; F52; G1
Parlog, E12, E22
partial differential equations, D19
pattern acquisition, F32
pattern matching, F32
pattern recognition, B36; D21; F55
PCP, E15
personal computer, C6, C16; D10; E37; F22
(See also microcomputer)
personal sequential inference machine, D31, D52; E3, E22
plant care, F62
plausible reasoning, C11; E5; F10, F39, F49, F67
Poplog, E31
POPS (OR-Parallel Optimizing Prolog System), E16
problem solving, A6, A10; B28, B33; F9, F15, F19, F27, F45, F47, F55, F56, F57
production system, F7, F39, F41, F42, F48
program synthesis, F11
program transformation, E35
Prolog, B23, B27, B42, B45; C8, C19; D20, D22, D23, D51; E1, E3, E4, E5, E7, E9, E11, E16, E17, E18, E20, E21, E22, E23, E25, E27, E28, E29, E31, E32, E36, E37; F24, F27, F46
(See also Concurrent Prolog, ESP, Micro-Prolog, Prolog/KR)
Prolog/KR, E25
Prospector, F22, F44, F45, F51, F63
Puff, F26

Qute, E28

R1, F28
reasoning, F32
(See also inference)
relational database, B3; C13; D5, D9, D23, D24, D30, D34; E21, E23; F55
Rita, F44
robotics, B31; F32

robots, B21
RX, F2

SAM, C10
scientific application, F5, F10, F42
Schlumberger Ltd, F1
search, C8; F9, F59
semantic net, F7, F39
SHRDLU, C10
signal processing, D3
SIM, E13
SIMPOS, E13
simulation, D19, D20, D21; E25, E32
SIS, F35
Smalltalk, C8
social context, A5, A10, A12; B16, B20, B21, B25, B33; F56; G2
social implications, B5; F1
software engineering, B1, B4, B5, B6, B15, B22, B24, B39
speech output, B45; C1, C4, C6; D3
speech understanding, B45; C7, C14, C16, C17, C18; D3; F9
(See also voice recognition)
SRC (Semiconductor Research Corporation), D55
Stanford Heuristic Programming Project, F2, F5. F26
supercomputers, B2, B5, B36, B44; D12
Supercomputers project, B22
SU/X, F26
symbolic computation, F35
Syracuse University, D20
System 5G, D33
systolic array, D25

Taxadvisor, F41
TDUS (Task-Oriented Dialogue Understanding System), C10
technology transfer, B37; F34
Teiresias, F26
Teknowledge Inc, F63
theorem proving, F11
theory of programming, B15
Tohoku University, B23
Tokyo Institute of Technology, B22, B23
Tokyo University, B22, B23
Toshiba Fuchu Works, B22
transputer, D36
Turing Institute, B31; F43

U-interpreter, D6
ultracomputer, G1
United Kingdom, A1, A4, A7, A8, A13; B1, B5, B7, B13, B14, B18, B31, B32, B35; D27; E22; F20, F43, F58
United States of America, B3, B4, B5, B10, B13, B14, B17, B18, B19, B25, B28, B37, B42; C18; D27, D46, D55; E22; F1, F40
University College, C5
user-friendly system, C9

vector processor, B36
(See also Cyber 205 computer, Cray computer)
Vili, C11

vision, A7; B5, B35; C2, C5, C11, C13; F32, F54
VLSI (Very Large Scale Integration), A1, A10; B1, B5, B7, B33, B40, B45; C1, C4; D2, D3, D4, D14, D16, D20, D28, D29, D32, D33, D38, D40, D41, D42, D43, D44, D45, D51, D54, D55; F2, F35; G1
voice recognition, B5, B21, B35; C1, C6, C15, C16; D56; F55
(See also speech understanding)

West Germany, B5; D27
Western Europe, B14; D46; F40
(See also specific countries)

Yokosuka Electrical Communciation Laboratory, B22